Routledge Revivals

The Marxist Philosophy and the Sciences

This book, first published in 1938, is based upon the Muirhead lectures on political philosophy delivered in the University of Birmingham in January and February of 1938. This title was intended to be of interest to students and scientific workers in the belief that Marxism will prove valuable to them in their scientific work, as well as to a wider audience.

AF867653

The Marxist Philosophy and the Sciences

J. B. S. Haldane

Routledge
Taylor & Francis Group

First published in 1938
by George Allen & Unwin Ltd

This edition first published in 2016 by Routledge
2 Park Square, Milton Park, Abingdon, Oxon, OX14 4RN
and by Routledge
711 Third Avenue, New York, NY 10017

Routledge is an imprint of the Taylor & Francis Group, an informa business

© 1938 J. B. S. Haldane

All rights reserved. No part of this book may be reprinted or reproduced or utilised in any form or by any electronic, mechanical, or other means, now known or hereafter invented, including photocopying and recording, or in any information storage or retrieval system, without permission in writing from the publishers.

Publisher's Note
The publisher has gone to great lengths to ensure the quality of this reprint but points out that some imperfections in the original copies may be apparent.

Disclaimer
The publisher has made every effort to trace copyright holders and welcomes correspondence from those they have been unable to contact.

A Library of Congress record exists under LC control number: 89006620

ISBN 13: 978-1-138-95473-1 (hbk)
ISBN 13: 978-1-315-66680-8 (ebk)
ISBN 13: 978-1-138-95474-8 (pbk)

J. B. S. HALDANE, F.R.S.

The Marxist Philosophy and the Sciences

LONDON

GEORGE ALLEN & UNWIN LTD

FIRST PUBLISHED IN 1938
SECOND IMPRESSION 1939
THIRD IMPRESSION 1942

All rights reserved

PRINTED BY THE REPLIKA PROCESS
IN GREAT BRITAIN BY
LUND HUMPHRIES
LONDON BRADFORD

Preface

THIS book is based upon the Muirhead lectures on political philosophy delivered in the University of Birmingham in January and February 1938. I have expanded them to deal more fully with matters of detail. They are primarily addressed to scientific workers and students, in the belief that Marxism will prove valuable to them in their scientific work, as it has to me in my own. But in view of the general importance of the subject I hope to interest a somewhat wider audience.

I have tried to apply Marxism to the scientific problems of my own day, as Engels did over many years, and Lenin in 1908. I do not doubt that I have made mistakes. A Marxist must not be too afraid of making mistakes.

Such an attempt as mine inevitably invites one of two criticisms. If one confines oneself to well-established scientific facts, one is told that it is easy to apply Marxism after the event, and that with sufficient ingenuity one can find a quotation from Marx or Engels which is apposite to any piece of recent scientific work. If, on the other hand, one ventures into speculation one is certain to be wrong on points of detail, if not on more fundamental matters. Nevertheless, I think it is worth while to demonstrate the kind of speculations into which Marxism leads a scientist.

For an acceptance of this philosophy inevitably induces novel types of action and thought. This must be my apology for parts of Chapter 5 which some readers will consider an excrescence on an otherwise useful book.

I have tried to cover a very wide field, and am fully aware that I have done so both unevenly and superficially. I hope, however, that I may stimulate others to fill in the gaps in my exposition and to correct it where necessary. An adequate Marxist interpretation of science can only arise in an atmosphere of vigorous controversy, and I wish to make it absolutely clear that I expect to be criticized. But I hope that the criticism to which I am subjected from Marxist writers will be constructive.

I have used the following abbreviations in my citations:

F. *Ludwig Feuerbach and the outcome of Classical German Philosophy*, Frederick Engels, 1888. English translation. (London, Martin Lawrence, N.D.)

A.D. *Herr Eugen Dühring's Revolution in Science (Anti-Dühring)*, Frederick Engels, 1878. English translation by Emile Burns. (London, Lawrence and Wishart, N.D.)

C. *Capital*, Karl Marx. English translation by Moore and Aveling, edited by Frederick Engels, 1887. (London, George Allen & Unwin Ltd.)

Preface

O.F. *The Origin of the Family, Private Property, and the State*, Frederick Engels, 1884. English translation by Ernest Untermann. (Kerr, Chicago, 1902.)

M.E. *Materialism and Empirio-Criticism*, V. I. Lenin, 1909. English translation by David Kvitko. (London, Martin Lawrence, N.D.)

I hope that many of my readers will be induced to study these books.

Contents

1. *Some Marxist Principles*

I OWE two apologies to my audience and readers. In the first place I am not primarily a philosopher; but when I am asked to lecture on political philosophy, I can choose no more appropriate subject than the most political of all philosophies, that of Marx. The second apology is more serious. I am by no means qualified to speak on Marxism. I have only been a Marxist for about a year. I have not yet read all the relevant literature, although I had of course read much of it before I became a Marxist. The object of these lectures will not only be to enlighten my audience, but to clarify my own thoughts. It will be remembered that Socrates described himself as the midwife who helped the unborn thoughts of others into the world. I will ask my hearers and readers to function in that capacity in my own case.

Now we must ask ourselves at once, why is Marxism important? I think I may presume that the majority of my audience and a considerable fraction of my readers is hostile to it. Why should they worry about it? One reason is because it is a philosophy of very great practical importance, a philosophy which is not less important if one decides that it is entirely false. It makes a considerable difference to the conduct of its adherents. I believe that one could spend a week (in vacation time, at any rate) with the average academic philosopher without discovering whether he was an idealist or a realist, but I do not think that one could spend a day with a Marxist without discovering his tenets.

There are two other important philosophies which issue in action to a very considerable extent. The first is the scholastic philosophy, whose greatest exponent was St. Thomas Aquinas. That philosophy represents not merely the opinions of a few people, or even of the whole body of priests and monks, but the practice of the great medieval civilization. That philosophy is still active in guiding the activity of the Roman Catholic Church. It is, therefore, deserving of study whether we adhere or object to it, simply because the Catholic Church is a very important institution. The second of these practically important philosophies is what a century or two ago was called natural philosophy, and is now called science. It is, however, limited in its scope. It has certainly been successful in some fields. In others it has had less application. It has undoubtedly transformed the world.

Now Marxism claims to apply scientific method in the field of politics and economics, and to predict and to enable us to control the transformation of the world still further. Because it extends scientific method into the human field it throws a new light on science, as a human activity depending both on contemporary social and economic conditions and also on certain very general laws of human thought. It further lays down some principles which are said to hold throughout nature, as well as applying to human activities. We shall have to investigate these claims in what follows.

Above all, I believe that I am justified in giving these lectures because of the very remarkable misapprehensions which undoubtedly exist in many quarters regarding the Marxist philosophy. A good many people do not, I think, even know of its existence. They know nothing of the theoretical side of Marx's work, except, perhaps, the doctrine of surplus value. If they hear that Marxism is materialism, they think materialism is the theory that man is a machine, or the denial of the existence of mind.

Now, until 1917, it might have been possible to dismiss Marxism as the doctrine of a small set of cranks, no more important than the doctrines of Bakunin, Sorel, or other revolutionary theorists. This was particularly so in England, where Marxism was largely ignored both in academic and political circles, whereas on the continent of Europe it was at least considered worthy of criticism. You will remember, however, one of the definitions of a crank, covering both the human and mechanical kinds, as "A little thing that makes revolutions"! It is now impossible to doubt the importance of Marxism, because Marxism was the philosophy of Lenin. It is very difficult to deny that Lenin was the greatest man of his time. Not that this admission need imply agreement with him. It is perfectly possible, without being a Mohammedan, to admit that Mohammed was the greatest man of his time. The philosophy of a man who has had so great and important an influence on world history as Lenin is undoubtedly worthy of investigation.

You will remember that Plato said that the ideal state was only possible when a philosopher became a king. Lenin was, amongst other things, a philosopher. We shall have to examine some of his philosophical views later on. He became, if not a king or even a dictator, the most important man, and the ideological leader, of a community covering most of the former Russian Empire. And that community is still largely guided by the principles which he laid down. The Soviet Union is certainly not the ideal state, for one reason because Marxists are not interested in ideal states, but in actual or possible states. Lenin's philosophy is today very much alive, both in the Soviet Union, and among communists and other Marxists who are not members of the communist party, outside the Soviet Union. The intensity of the interest taken in philosophy in the Soviet Union may be gauged by the statement, which I believe to be true, that in 1936, one hundred thousand copies of a translation of certain of Kant's works (I cannot believe they were his complete works!) were printed, and the whole lot sold out. Philosophy is at any rate a subject of very general interest in the U.S.S.R., and one result of communist propaganda in Britain has been a revival of interest in philosophy.

Such are some of the reasons why even those who are convinced of the truth of some other philosophy, and of the rightness of some other political practice, should be willing to make at least a superficial study of Marxism. My own reason for delivering these lectures is a different one. I think that Marxism is true.

Some Marxist Principles

Now, what is Marxism? Plekhanov, a Russian Marxist and predecessor of Lenin, began his book, *Fundamental Problems of Marxism,* with the statement: "Marxism is a complete theoretical system." That is approximately true of the philosophy of Aristotle, St. Thomas, Spinoza, or Hegel. Clearly it is not true of the philosophy of Socrates. It is also untrue of Marxism. Marxism is not complete, not a system, and only in the second place theoretical. It is not complete because it is alive and growing, and above all because it lays no claim to finality. The most that a Marxist can say for Marxism is that it is the best and truest philosophy that could have been produced under the social conditions of the mid-nineteenth century. It is not primarily a system, but a method. As Marx said in the Eleventh Thesis on Feuerbach:* "The philosophers have only interpreted the world in various ways: the point is to change it." Like Descartes, he regarded his philosophy as primarily a method, and although theory is essential in Marxism, Marx proclaimed the primacy of practice over theory.

This is not, of course, to say that Marxism does not include a great deal of systematic theory, which is to a large extent the fruit of the method. But the details of Marxist theory, like those of the theories of natural science, are the result of applying the method to concrete situations. And the theory which exists was built up with far more attention to observed facts and far less "pure thought" than the great philosophies of the past.

A few words are necessary about the historical origins and sources of Marxism. Marx was born in 1818 at Trier, in south-western Germany; his father was a Jewish

* F., p. 75

lawyer who became a Protestant when Karl was six years old. His colleague, Friedrich Engels, to whom Marxism owes so much, was born in 1820 at Barmen, in Rhineland. His father was a German manufacturer. Both studied philosophy. Marx got his Doctorate for a thesis on the philosophy of Epicurus; they both became left-wing Hegelians, and later followers of Feuerbach. Marx wanted to become a philosopher, and it is likely that had he become a professor he would have been a good deal more innocuous to the social system in which he lived than actually proved to be the case. However, the Prussian Government dismissed a number of people like Feuerbach and Bauer, whose philosophical and political views, though radical, were very much milder than Marx's views later became. Marx took to journalism. He was one of the founders of the radical *Rheinische Zeitung* in 1842. When it was suppressed in 1843, he went to Paris. Meanwhile, Engels had gone to Manchester in 1842, where he worked as a cotton broker and studied the life of the working people. His book, *The Condition of the Working Class in England*, was published in 1845.

In 1844, Marx and Engels met in Paris, and became lifelong friends. They came under the influence of French revolutionary theorists like Proudhon, and became Socialists. In 1845, at the instance of the Prussian Government, Marx was forced to leave Paris, and went to Brussels. By this time their views had been considerably clarified. They were in disagreement with Proudhon and other French leaders, and their economic and political outlooks were stated in the "Communist Manifesto," which Engels drafted, and which was published in 1847.

In 1848, both took part in the revolution in Germany, Marx as a journalist, Engels as a soldier. From 1849 onward, they lived most of their lives in England until Marx died in 1883 and Engels in 1895. Marx lived in London, Engels in Manchester until 1871, when he, too, came to London.

Apart from their political work such as that involved in founding the International Working Men's Association, later known as the First International, they wrote on a very large scale during those years. Marx's most important book was, of course, *Das Kapital,* but in discussing the relation of Marxism to science we shall mainly be concerned with the views expressed by Engels. His most important books for our purpose are, *Herr Eugen Dühring's Revolution in Science,* written in 1878, and popularly known as *Anti-Dühring,* and a smaller book called *Ludwig Feuerbach and the Outcome of Classical German Philosophy,* written in 1888. Finally, we have a large number of manuscript notes of Engels, which although never published in book form, have appeared in the Marx-Engels Archiv, under the title "Dialektik und Natur." *Anti-Dühring* and *Feuerbach* are both polemical works, and most people find them easy to read. But they are somewhat of a puzzle for an ordinary philosophical student, for a number of reasons. It is important to remember that Dühring, whose writings Engels analysed with considerable sarcasm,was a Socialist and a materialist, with whom he had enough in common to furnish a real basis of argument. He attacked him with the greatest vehemence on the points on which they differed.

Again, Engels professed himself a disciple of Feuerbach, but was critical of his opinions in a number of respects. Similarly, his joint work with Marx, *The Holy Family*, was directed against Bruno Bauer, with whom they were in a considerable measure of agreement, and the *Poverty of Philosophy* was directed against Proudhon. It is a characteristic of all these books that they are written, not against open enemies, but against persons with whom the authors had a good deal in common. This makes them difficult reading for one who is accustomed to the average philosophical work, which is addressed to the whole world, so to speak, and not to a group with which the author has only a limited number of bones to pick.

Why, it may be asked, should Engels not have attacked such contemporaries as Comte, Mill, Spencer, or Green, from whom he differed on almost all points? Perhaps the answer is as follows. It was obvious that such philosophies as theirs would become obsolete in a relatively short time. Many of the political and economic theories of Mill and Spencer are simply irrelevant to modern conditions. On the other hand, the views of Dühring and Feuerbach are held by a good many modern Socialists. Engels attacked "right" theories, not in their crude form, but in their most dangerous form. In fact, he chose not the easiest, but the most difficult antagonists.

Lenin's only important philosophical work is called *Materialism and Empirio-Criticism.* It was published in 1908 and was directed mainly against Bogdanov, Lunacharsky, and others who claimed to be Marxists. Lunacharsky later became one of Lenin's colleagues in the Government of the U.S.S.R. At first reading, Lenin's book might seem to be an attempt to impose a formal, narrow

Marxist orthodoxy. Actually he is undoubtedly justified, when Bogdanov claims to be a Marxist, in quoting passages from Marx which disagree with Bogdanov. The whole book is a book characteristic of a fighter. His attacks are mainly directed against compromisers, whether within the Marxist movement, like Bogdanov and Lunacharsky, or outside it, like Mach and Avenarius. His opinion was that "non-partisans in philosophy are just as muddle-headed as in politics." On the other hand he recognized and admired clear thinking wherever he found it, and in consequence was often extremely polite to his out and out opponents, like James Ward, to whom he refers in the following sentence, among others: "The question, as put by this frank and consistent spiritualist is remarkably clear and to the point." Similarly, Karl Pearson is described as "this conscientious and honest enemy of materialism." Besides this book, some short but most important manuscript notes by Lenin on philosophical problems have been published.

In these lectures, we shall mainly be concerned with the relationship of Marxism to science, as developed by Engels in *Feuerbach* and *Anti-Dühring*, and by Lenin. Lenin's welcome to the new developments in physics, such as radio-activity and electrons, is particularly interesting as showing the relation of Marxism to discoveries which have been supposed to disprove its basic principles. However, Engels is the chief source, although he states expressly that most of the leading principles in his work derived from Marx.

Now a student of academic philosophy who takes up a study of Marxism will at first be disappointed. A great many questions are left unanswered, for two different reasons. Some were shown to be improperly put, and it was sufficient to demonstrate the historical reasons why they had been asked in the past. Others could not be answered on the existing data. Thus the relation between brain and mind is not in principle an insoluble problem; but it cannot be solved, except in the most summary manner, until we know a very great deal more, particularly about the brain. Marxism is not concerned mainly with being, but with becoming. It claims to enable us to understand change and development of all kinds, not only political and economic change and development, and by understanding to influence and to control them.

Most philosophers have treated time and change as more or less illusory, though since Hegel's day they are more often taken seriously. An attempt is made to find a timeless being behind this changeable world. That is conspicuously so in the philosophy of Plato. It is worth pointing out that Christianity differs from most of the academic philosophies in ascribing a supreme importance to a number of events in time—the Creation, the Fall, the Redemption of Mankind, and the Last Judgement. That was particularly so in primitive Christianity; and as it ceased to be a revolutionary religion, certain theologians tried to make its theory more and more static. In the first centuries of Christianity, theology was considerably influenced by the neo-Platonists, and in our own day we find such philosophers as Dean Inge trying to minimize the temporal side of theology and to exalt the timeless side. It is not, of course, a mere coincidence that their political views are usually reactionary.

While Marxism makes what at the very least must be admitted to be an ambitious attempt to solve the problems of becoming, it has very little to say concerning the problems of being raised in the classical philosophies. It dismisses many of them as illusory problems which have arisen through unclear thinking. It postulates nothing behind matter, and therefore dismisses metaphysics. It certainly postulates an inexhaustible supply of properties of matter, but no more than that.

In the remainder of this chapter, I shall try to summarize some of the principles of Marxism, though mainly outside the economic field. I shall only deal in the most summary way with some of Marx's economic and political theories in the last chapter; and as I am not an economist, I do not pretend that my treatment will be either novel or authoritative.

In the first place, we have the principle of the unity of theory and practice, with the primacy of practice. Let me quote one of Marx's theses on Feuerbach:*

> "The question whether objective truth can be attributed to human thinking is not a question of theory but is a practical question. In practice man must prove the truth, i.e. the reality and power, the 'this-sidedness' of his thinking. The dispute over the reality or non-reality of thinking which is isolated from practice is a purely scholastic question."

Again, Engels, in writing of astronomy, pointed out that in its early days, the Copernican hypothesis was only one of a number of theories, each of which would explain the facts with sufficient accuracy. It was only when, on the basis of Newton's theory of gravitation it was possible to predict such events as the finding of the planet Neptune and the return of Halley's Comet, that it could

* F., p. 73.

be taken as proved. All this is a commonplace for modern scientific theory, but it was by no means commonplace ninety years ago.

So far we may say that Marxism anticipates pragmatism, although it differs from pragmatism in almost all other respects, notably in its consistent emphasis on the changing of the world, and above all in its belief that there is a real world, and that absolute truth, if never reached, can be continually approached.

A second Marxist principle is materialism. This word has been used in a very large number of senses, and it is important to realize just what Marx meant by the term. Engels wrote* as follows:

"The question of the relation of thinking to being, the relation of spirit to nature—the paramount question of the whole of philosophy—has, no less than all religion, its roots in the narrow-minded and ignorant notions of savagery. But this question could for the first time be put forward in its whole acuteness, could achieve its full significance, only after European society had awakened from the long hibernation of the Christian Middle Ages. The question of the position of thinking in relation to being, a question which, by the way, had played a great part also in the scholasticism of the Middle Ages, the question: which is primary, spirit or nature—that question, in relation to the Church, was sharpened into this: 'Did God create the world or has the world been in existence eternally?'

"The answers which the philosophers gave to this question split them into two great camps. Those who asserted the primacy of spirit to nature and, therefore, in the last instance, assumed world creation in some form or another—(and among the philosophers, Hegel, for example, this creation often becomes still more intricate and impossible than in Christianity)—comprised the camp of idealism. The

* F., p. 31.

others, who regarded nature as primary, belong to the various schools of materialism.

"These two expressions, idealism and materialism, primarily signify nothing more than this; and here also they are not used in any other sense."

You will notice the emphasis which is laid on temporal priority rather than on logical priority. This is characteristic of a philosophical system which takes historical fact extremely seriously.

Now that definition of materialism is not accepted by many people. For example, my late father, J. S. Haldane,* wrote as follows:

"Materialism may be defined as the belief that physico-chemical realism, or the assumption that the representation of our surrounding universe by the physical sciences in their traditional form corresponds to reality, can be extended so as to cover, not only the phenomena of life, but also those of conscious behaviour."

If we compare this with what Lenin wrote, we shall see that J. S. Haldane's view, so far at least as it is expressed in that passage, is not in conflict with Marxism. Lenin's words are:

"It is, of course, totally absurd that materialism should maintain the 'lesser' reality of consciousness or should necessarily adhere to a 'mechanistic world-picture' of matter in motion and not an electro-magnetic, or even some immeasurably more complicated one."†

Again, in another place, Lenin wrote:

"For the sole 'property' of matter—with the recognition of which materialism is vitally concerned—is the property

* *Materialism*, p. 5. (London, 1932.)

† M.E., p. 238.

of being *objective reality*, of existing outside of our cognition. . . . The recognition of immutable elements, 'of the immutable substance of things,' is not materialism, but is metaphysical, anti-dialectical materialism."*

It is clear, therefore, that what Marxism calls materialism is something a good deal less mechanical than the materialism of the French eighteenth-century philosophers. It is worth noting that although my late father was a strong opponent of materialism, his book, *The Sciences and Philosophy*, was recommended by a Moscow radio commentator as a very good introduction to dialectical materialism, although far from being Marxist.

Again, Lenin's attitude to idealism, although hostile, was not completely negative:

"Philosophical idealism is nonsense only from the standpoint of a crude, simple, and metaphysical materialism. On the contrary, from the standpoint of dialectical materialism philosophical idealism is a one-sided, exaggerated, swollen development (Dietzgen) of one of the characteristic aspects or limits of knowledge into a deified absolute, into something *dissevered* from matter, from nature."†

Why, then, many people ask, should you not drop this word "materialism" which has come to signify addiction to large dinners and expensive motor cars, and call it "realism" or something less challenging? The answer is that Marxism insists on the priority of matter, and that it is a fighting philosophy. Marxists must on occasion deal very vigorously with idealists. Today, for example, it is necessary to combat the propaganda of those pacifists who believe the world can be saved from war by goodwill acting, as it were, in a vacuum, and with the

* M.E., p. 220.

† M.E., p. 327 (manuscript notes).

anarchists, who think that it is sufficient to destroy the existing State organizations, and human nature is good enough to do the rest.

The materialism of Marxists is called dialectical materialism, for a reason which will be explained later. Dialectical materialism as applied to human history is called historical materialism. This is the aspect of Marxist philosophy which is probably most familiar to British readers. But it is only one aspect, and is not that with which we shall be mainly concerned in this book. The nature of the materialistic interpretation of history will be made clear by two quotations from Engels:

"The new facts made imperative a new examination of all past history, and then it was seen that all past history was the history of class struggles, that these warring classes of society are always the product of modes of production and exchange, in a word, of the economic conditions of their time; that therefore the economic structure of society always forms the real basis from which, in the last analysis, is to be explained the whole superstructure of legal and political institutions, as well as of the religious, philosophical, and other conceptions of each historical period. Now idealism was driven from its last refuge, the philosophy of history; now a materialistic conception of history was propounded, and the way found to explain man's consciousness by his being, instead of, as heretofore, his being by his consciousness."*

Again, in another passage he says: †

"In modern history at least it is therefore proved that all political struggles are class struggles, and all class struggles for emancipation in the last resort, despite their necessarily political form (for every class struggle is a political struggle), turn ultimately on the question of economic emancipation.

* A.D., p. 32.

† F., p. 62.

Therefore, here at least, the state, the political order, is the subordinate, and civil society—the realm of economic relations—the decisive element. The traditional conception, to which Hegel, too, pays homage, saw in the State the determining element, and in civil society the element determined by it. Appearances correspond to this. As all the driving forces of the actions of any individual person must pass through his brain, and transform themselves into motives of his will in order to set him into action, so also all the needs of civil society—no matter which class happens to be the ruling one—must pass through the will of the State in order to secure general validity in the form of laws. That is the formal aspect of the matter—the one which is self-evident. The question arises, however, what is the content of this merely formal will—of the individual as well as of the State—and whence is this content derived? Why is just this intended and not something else? If we enquire into this we discover that in modern history the will of the State is, on the whole, determined by the changing needs of civil society, by the supremacy of this or that class, in the last resort, by the development of the productive forces and relations of exchange.[11]

The detailed economic theories of Marxism lie outside the scope of this book, as does the description of Marxist practice in the present class struggle.

Dialectical materialism is founded on Hegelian dialectic. It had long been realized that matter on the whole behaves intelligibly, conforming to the laws of logic and arithmetic. The question arose whether our reason mirrors the behaviour of matter, or whether on the other hand, matter mirrors the behaviour of mind. Kant's view was somewhere intermediate, perhaps leaning to the idealist side.

Hegel laid down, especially in his *Logic* and *Phenomenology of Mind*, a number of principles of thought going beyond those laid down by Aristotle and taught as formal logic, principles which had been more or less recognized for centuries, but never so clearly formulated. These principles were called dialectical principles. He said that nature conformed to them. According to Hegel the logical categories exist eternally; the world is a mere exemplification of these logical categories in space and time. Feuerbach, Marx, and Engels believed that the principles were exemplified in nature before they governed thought. According to Marx, the ideal is nothing but the material world reflected by the human mind, and translated into forms of thought. Hegel is standing on his head. Our business is to put him on his feet. Engels treated the Hegelian dialectic as expressing primarily the properties of matter, and only secondarily the laws of thought. He held that the principles which Hegel had worked out in the realm of thought also applied to material events, not only in the social field, but in the fields of astronomy, physics, biology, and so on.

In what follows I propose to give a sketch of the dialectic so brief and abstract as to be almost a caricature. I shall pass over many of its essential features, and attempt to summarize a few of its main principles. Such a presentation lays itself open to a severe criticism. The dialectic, which is a unity, appears as a collection of rules of thumb, one or other of which should be applied wherever possible. Such a point of view would, I am sure, be dispelled by a reading of Marx, Engels, and Lenin, or of a more modern exposition such as Jackson's *Dialectics*. I hope, indeed that the following chapters may serve to dispel it. If not, the fault is my own, not that of Marx, Engels, and Lenin.

What are these dialectical principles? One of them is the principle of the unity of opposites. For example, if I say, "John Smith is a man," I am asserting the identity in a certain context of a particular, John Smith, and the universal, man. This identity has led philosophers into very great difficulties for the last 2,300 years. Again, I say that the wood of which this table is made is hard, or it would not support things, and soft, or it could not be cut. Two opposite qualities are united. Before such assertions, we have two alternatives; we may say, as Plato said, that matter is something self-contradictory, it is and is not. Universals are real, but matter is unreal.

Or we may say with Engels that matter unites these opposites. This means that matter is something very much richer and more complicated than the mechanistic materialists had ever dreamed.

Two remarks may be made on this principle. Lenin* wrote that the unity of opposites is something conditional and temporary. Gas has no hardness, in the sense that it will put up no permanent resistance to division. On the other hand, it is probable that an electron is absolutely hard in the sense of being completely indivisible.

At any stage in the development of science we can undoubtedly explain away contradictions which puzzled our ancestors. For example, today, instead of saying, like Plato, that a table is both hard and soft, we can ascertain by a number of measurements the degree of hardness of the wood, its breaking strain, and so on.

There are a number of things which were paradoxical

* M.E., p. 324.

to Plato and are not to us. On the other hand, in our own time new contradictions have appeared which seem just as trying to us as contradictions which we find trivial appeared to Plato. For example, electrons have apparently at the same time properties which compel us to regard them as particles, and other properties which can be explained if they are systems of waves. Two thousand years from now, these difficulties may seem very elementary indeed, but I think that our descendants will probably still be finding opposites embodied in matter which they will find difficult to unify.

The second principle is the passage of quantity into quality, and conversely. This phrase is taken from Hegel, but a much more satisfactory account of what is meant by it is given by Marx* in *Capital.*

> "Here, as in natural science, is verified the correctness of the law discovered by Hegel in his 'Logic' that merely quantitative changes beyond a certain point pass into qualitative differences."

A classical example of this is the boiling or freezing of water, but any other change of phase in physical chemistry may be taken as an example. At the boiling point of water some of its measurable qualities show an abrupt break. The volume, which has been going up steadily but slowly, shoots up enormously. Other properties disappear; for example, the capacity for dissolving solutes and that of ionizing salts.

The principle is, of course, absolutely fundamental in physiology. A hundred years ago it was commonly said that carbon dioxide was a poison, because a man died if he breathed pure carbon dioxide. Then J. S. Haldane

* C. (I), p. 336.

found that a certain amount of this substance was essential for life. The normal amount in the blood corresponds to a pressure of about 5 per cent of an atmosphere. If this is either doubled or halved serious symptoms arise. In fact, too much carbon dioxide is a poison, but a certain amount is a necessity.

It is equally fundamental in such ethical systems as that of Aristotle, who pointed out that the difference between good and evil was largely quantitative. Thus the coward took too few risks, the rash man too many, and the brave man the right number. It is even familiar in law, where, for example, three or more people can make a riot, but one or two cannot.

In modern physics it is familiar under the name quantization. Not only mass, but energy, can only be transferred from one system to another (at least in certain cases) in definite quantities. We shall deal with this matter in more detail in Chapter 3. It may well be that quantum phenomena are the most fundamental and primitive expression of this principle, and that the other examples of it will ultimately be explicable on a basis of quantum theory.

Now according to the view of matter which was first clearly formulated by Locke, though it goes back to Descartes and Democritus, the quantitative aspect of matter is real, whilst many of its qualities are illusory. Thus what we call colours and tones are "really" only vibration frequencies. But for Marxists both quantity and quality are properties of the real world.

The converse transformation of quality into quantity is exemplified when a symphony is recorded on a sound-track. Since our knowledge of the external world depends on the frequencies with which nervous impulses reach our brains and spinal columns along a million or two nerve fibres, and not on qualitative differences in these impulses, this transformation, and the reverse transformation of quantity into quality which takes place in our brains, play a fundamental part in our knowledge of the world.

The social applications of the principle are important. They have been given by Marx in a number of places, and it is worth while pointing out that laws holding right through one state of society may become meaningless in another. Social change may be discontinuous, as in the case of water to steam at atmospheric pressure, or continuous, as in the case of the passage from water to steam at pressures higher than the critical pressure. Thus slavery came to an abrupt and violent end in the United States, but faded out gradually in the late Roman Empire.

A third principle, which is perhaps the most important, is what is called the negation of the negation. Let me give a simple example. I learn to drive a motor car, and among other things to steer it. Then I drive a little faster than usual, and skid. Skidding is the negation of steering. After skidding a number of times, I learn to control a skid in the direction which I desire. That is a passage to a higher level of motor driving. It is a passage which some drivers never make. London bus drivers, who have to learn to drive in pools of oil, are compelled to make it, and the controlled skid is part of the technique of every racing motorist. I take that example from familiar practice. Examples in physics and biology will follow later on.

One of Marx's examples from economics is interesting as showing how he applied this idea in the economic field. First of all, he describes medieval English industry, in which the workers owned the means of production, their own tools, and, in some cases, their own land; but he was particularly concerned with handicraft production. Then, with the development of industry in the early stages of capitalism, the immediate producers were expropriated, ceasing to own their means of production, either forcibly, as through the enclosures of the land, or more generally by the competition of far more efficient industry based on division of labour and on capital. The hand-looms were killed by the factories. This process was the negation of the ownership by the workers of their means of production. But Marx claims that this process is now being negated. In the present stage of capitalism, capital is negating itself.*

"That which is now to be expropriated is no longer the labourer working for himself, but the capitalist exploiting many labourers. This expropriation is accomplished by the action of the immanent laws of capitalistic production itself, by the centralization of capital. One capitalist always kills many. Hand in hand with this centralization, or this expropriation of many capitalists by few, develop, on an ever extending scale, the co-operative form of the labour process, the conscious technical application of science, the methodical cultivation of the soil, the transformation of the instruments of labour into instruments of labour only usable in common, the economizing of all means of production by their use as the means of production of combined socialized labour. . . . Along with the constantly diminishing number of the magnates of capital, who usurp and monopolize all advantages of this process of transformation, grows the

* C. (I), p. 386.

mass of misery, oppression, slavery, degradation, exploitation; but with this too grows the revolt of the working class, a class always increasing in number, and disciplined, united, organized by the very mechanism of the process of capitalist production itself. The monopoly of capital becomes a fetter upon the mode of production, which has sprung up and flourished along with, and under it. Centralization of the means of production and socialization of labour at last reach a point where they become incompatible with their capitalist integument. This integument is burst asunder The knell of capitalistic private property sounds. The expropriators are expropriated."

That is the way in which Marx conceived of the development of capitalism, its self-destruction, and the coming of socialism.

Now the negation of the negation was regarded by Marx as the main source of progress and of novelty. A great many philosophers, for example, Lloyd Morgan and Smuts, have recently been interested in what they call the emergence of novelty.

Lenin wrote:

"Two fundamental (or is it the two possible? or is it the two historically observed?) conceptions of evolution are: development as decrease and increase, as repetition; and development as a unity of opposites (the division of the one into mutually exclusive opposites and their reciprocal correlation). The first conception is dead, poor, and dry; the second is vital. It is only this second conception which offers the key to understanding the 'self-movement' of everything in existence; it alone offers the key to understanding 'leaps,' to the 'interruption of gradual succession,' to the 'transformation into the opposite,' to the destruction of the old and the appearance of the new."*

* M.E., p. 323 (manuscript notes).

We shall have to see how far this rather ambitious claim can be verified in the field of science.

Above all, dialectical materialism insists on the reality of change. It claims to go back beyond Plato and Socrates to Heraclitus, and in particular it welcomed the new developments of physics which seemed to some to spell the end of materialism, and which undoubtedly disproved the very narrow forms of materialism current in many scientific circles at the end of the nineteenth century, and still current in some of them.

As we saw above, Lenin wrote:*

> "The sole 'property' of matter—with the recognition of which materialism is vitally connected—is the property of being objective reality, of existing outside our cognition"

and for that reason he was very far from upset by the revolutionary physical discoveries of his time.

Again, Engels† said:

> "The great basic thought that the world is not to be comprehended as a complex of ready-made things, but as a complex of processes, in which the things apparently stable no less than their mind-images in our heads, the concepts, go through an uninterrupted change of coming into being and passing away, in which, in spite of all seeming accidents and all temporary retrogression, a progressive development asserts itself in the end—this great fundamental thought has, especially since the time of Hegel, so thoroughly permeated ordinary consciousness that in this generality it is scarcely ever contradicted. But to acknowledge this fundamental thought in words and to apply it in reality in detail to each domain of investigation are two different things."

* M.E., p. 220.

† F., p. 54.

You will see that in the idea of process as fundamental, we have the anticipation of much of what is valuable in the philosophies of Bergson and Whitehead. Later on I hope to show how these principles work, or at least to examine whether they work, in the field of science.

I am perfectly aware that my approach has been extremely incomplete. If anyone wishes to study the matter in detail, I would recommend them to read *Feuerbach* and *Anti-Dühring*, remembering that they were written from the point of view of the science of sixty years ago, and that therefore certain of the statements made in them would obviously have to be modified to meet recent developments of science.

An important type of dialectical process is as follows. We study some thing or some process in isolation. We produce a theory and we find that that theory is unsatisfactory because we have ignored the background. Now afterwards it is very easy for any critic to say, "Well, your original theory was just a piece of absurdity. Anyone could tell that it wasn't going to work!" Unfortunately, in practice we find that until we had produced the theory which worked up to a point and then broke down, we could not tell what elements we had ignored and should not have ignored. Let us take an example from chemistry. In the Middle Ages no self-respecting alchemist would have dreamed of doing any chemical process which was in any way difficult without first observing the position of the planets. For example, if it was an operation involving tin, he would presumably have seen that the experiment was begun when Jupiter was in the ascendant, because Jupiter was the planet presiding over tin. One of the greatest steps in chemical progress ever taken was when some bold man actually began making experiments without first observing the planets, and found that they were just as successful as before. Nevertheless, when chemical theory and practice progressed, it was found that there were certain things in the background which could not be ignored, things of which the medieval alchemists had never even dreamed. For example, it is clear that in any chemical experiment involving the measurement of the amount of gas produced, it is necessary to read not only the thermometer, but the barometer; and it is only when one takes account of the variations in the barometric pressure that one gets anything like exactitude in such measurements.

Now this increasing importance of the background is often a part of the historical process. For example, in what we believe to be the most primitive type of human life known to us, the collecting stage, which comes even before the stage of hunting, it is clear that the most effective community was the family. The same is still true of a society based on very simple hunting and fishing. In that stage of human development, the only sensible philosophy was anarchism—let your neighbour alone. Larger communities, however, are necessitated by denser populations due to more effective production, and some form of organization above the family becomes necessary. You can no longer neglect the background of other human families.

In the same way it is believed by many people that whereas a hundred years ago the national state could be regarded to a very considerable extent in isolation, this is now no longer possible, owing to the great development of transport, including the transport not only of men and merchandise, but also of bombs.

There is a special case which arises when the situation is altered by our own knowledge of it. Engels attributed to Hegel the statement: "Freedom is the recognition of necessity." I think that actually the first man who made that statement was not Hegel, but Spinoza. It is a paradox, but in many cases it is true. Let us take the following statement: "If you drink water polluted with *Bacillus typhosus*, you will probably get typhoid fever." That statement is substantially true, until we recognize that it is true, and take action based upon it. Until its truth was recognized, men tried all sorts of methods of dealing with typhoid epidemics; magic, prayer, war on bad smells, and so on, without very conspicuous success. Now the curious thing is that when that statement regarding typhoid was not only put forward, but was made the basis of action, it ceased to be true. It immediately became a lie, because you have to add to the words "you will probably get typhoid fever," "but not if your water is boiled or chlorinated, or if you get yourself immunized." In other words, by recognizing the necessity, you are able in that case and in many others to circumvent it.

The same thing is true of the doctrine of historical materialism. It may be claimed, in my opinion with a very large measure of truth, that man is to a considerable extent a slave of economic conditions, until he recognizes the fact; and the idealist, who denies the principle of historical materialism completely, is as much in the grip of economic conditions as anyone else. Marxists believe that the principle of economic determinism of other human activities is largely true, but they are out to make it untrue by founding a society in which economic classes have been abolished, and in which this particular kind of determinism no longer holds.

Of course, no Marxist would claim that before Marx's time no one struggled against economic conditions. On the contrary, almost if not quite all the political struggles of the past were at bottom struggles against economic conditions. The struggle was often unconscious, but sometimes fully conscious. But the participants in these past struggles concentrated on their immediate problems, and did not see them in their full historical perspective. The fact that Marxism lays so much stress on this struggle of human beings against economic forces makes it clear that the doctrine of economic fatalism is no part of the Marxist philosophy. On the contrary, Marxism unifies the theory of the struggle against economic fatalism with its actual practice.

The above is a very characteristic type of dialectical process, on which Professor Levy has laid particular stress in a number of papers and a recent book.*

Before we pass on, I want to compare this Hegelian-Marxist dialectic for one moment with that of Socrates, who may be said to have introduced the dialectical method into philosophy.

So far as we can make out, the Socratic method of operation was as follows: he started a conversation with some unfortunate Athenian citizen on a topic such as the nature of justice, and made his unlucky and unsuspecting interlocutor contradict himself. As a result of those contradictions, he arrived, if not at the truth, at any rate somewhat nearer the truth than his starting-point.

* *A Philosophy for a Modern Man*, Gollancz, 1938.

Some Marxist Principles

Plato wrote that the dialectical method was a means of arriving at absolute truth. For example, if the question discussed was, "What is Justice?" Plato thought that justice corresponded with some eternal idea, and that by examining the ordinary man's idea of justice, showing where it contradicted itself, and in consequence amending it so that it was no longer self-contradictory, he could arrive at a knowledge of that eternal idea of justice. We now, most of us, doubt whether Plato was correct; and there has been a tendency, especially perhaps among scientific people, to say that Socrates was merely investigating the meaning of words, and doing something pretty unimportant. I believe that this view is also incorrect. The word "justice" in Athens stood, if not for an eternal idea, at any rate for a social reality for which men were willing to die or to kill. But justice in Athens, even justice as conceived by the most enlightened Athenian, was by no means the same as justice in England today. Very few, if any, of Socrates' interlocutors would have regarded slavery as an essentially unjust institution; and in the same way, justice in twentieth-century England presumably means something different from what it will mean a hundred years hence.

We may conclude then, that while this verbal or argumentative dialectical process can take us a certain way, can clarify our ideas to a considerable extent, yet history applies a dialectical process of a far more searching character to our social institutions, bringing out contradictions which no amount of mere argument would have disclosed.

The Marxist theory of truth is, I think, straightforward and simple, but by no means complete. The view taken is that an indefinite progress is made in the direction of truth, except, perhaps, on fairly trivial matters such as the date of a given man's birth or death. This doctrine is of course, familiar to English students of philosophy in a slightly different form in the work of Bradley. A short quotation from Engels* states the Marxist point of view clearly:

"The sovereignty of thought is realized in a series of extremely unsovereignly-thinking human beings; the knowledge which has an unconditional claim to truth is realized in a series of relative errors; neither the one nor the other can be fully realized except through an endless eternity of human existence.

"Here once again we find the same contradiction as we found above, between the character of human thought, necessarily conceived as absolute, and its reality in individual human beings with their extremely limited thought. This is a contradiction which can only be solved in the infinite progression, of what is for us, at least from a practical standpoint, the endless succession, of generations of mankind. In this sense human thought is just as much sovereign as not sovereign, and its capacity for knowledge just as much limited as unlimited. It is sovereign and unlimited in its disposition, its vocation, its possibilities and its historical purpose; it is not sovereign and it is limited in its individual expression and in its realization at each particular moment.

"It is just the same with eternal truths. If mankind ever reached the stage at which it could only work with eternal truths, with conclusions which possess sovereign validity and have an unconditional claim to truth, it would then have reached the point where the infinity of the intellectual world both in its actuality and in its potentiality had been exhausted, and this would mean that the famous miracle of the infinite series which has been counted would have been performed."

* A.D., p. 101.

On the whole we may take it that Marxists are rather sceptical of the more ambitious logical theories. For example, the system of Russell and Whitehead in the *Principia Mathematica* is doubtless true, or largely true, if sufficiently sharp classification is possible.

It is, of course, based on the hypothesis that existents (e.g. dogs, lightning flashes, and sensations), relations (e.g. greater than, father of, desired by), and propositions (e.g. this hat is black, all pigs have heads, I want a drink), can be arranged in classes. Then, for example, if a one to one correspondence can be made between the members of two classes, say the bright stars in the Plough and the deadly sins of catholic theology, these two classes are members of a class which also includes the class of days in the week and the class of dwarfs who befriended Snow-White. This super-class is the number seven. And on this basis the fundamental theorems of mathematics can be proved.

If then we can divide up all animals precisely into different species, between which the distinctions are at all times well marked, no doubt the Russell-Whitehead theory of classification will hold. But actually this division of animals into species or other higher categories is by no means universally valid. The gap between species is bridged not only by evolution in the past, but, in some cases at any rate, by hybridization in the present. Engels made very great play with animals which bridged gaps—*Archaeopteryx*, which bridged the gap between reptiles and birds, and *Ceratodus*, bridging to some extent the gap between the fish and the amphibia, though, of course, far less completely than many fossil forms since discovered. For that reason it is probable that too great emphasis has been attached to logical systems which will only work for material that has certain highly abstract properties, which are rather less frequently and much less completely exemplified in the real world than logicians would like to think.

2. *Mathematics and Cosmology*

In the application of Marxism to science, we must proceed with the very greatest caution. At best Marxism will only tell a scientist what to look for. It will rarely, if ever, tell him what he is going to find, and if it is going to be made a dogma, it is worse than useless.

In the first ten years of scientific research in the U.S.S.R. certain writers—one can hardly dignify them by the title of workers—attempted to apply dialectical materialism to every kind of activity from portrait painting to fishing. They produced a great deal of utter nonsense. Indeed, the worse a scientific paper was, the more likely it was to be embellished with irrelevant quotations from Marx, Engels, and Lenin. Good science needs no such justification, and an experienced Marxist will notice evidences of dialectical thinking without any need to draw his attention to them. The curious reader will find a number of ludicrous examples of bogus dialectical materialism in an article by Stetski in the *Pravda* of June 4, 1932, which has been extensively cited in other countries to prove the decay of science in the Soviet Union. He may forget that Stetski, a member of the central committee of the communist party, had deliberately picked the most ridiculous out of many thousands of scientific and technical articles in order to check a dangerous tendency.

If dialectical materialism has sometimes been misapplied in the U.S.S.R. this does not imply that a little more of it would hurt British scientific thinkers. It is, in fact, already being used by a small number of them. Unfortunately a citation of some of the best examples might endanger the posts and salaries of the workers concerned.

Most scientific work is done in a limited field. It no more needs dialectic than it needs the differential calculus or a microscope. Nevertheless, I am convinced that Marxism proves of the greatest value in studying the development of science, and the relationship of the different sciences to one another, particularly the relation of chemistry to physics, and of biology to chemistry. And it is particularly useful in those branches of science which are themselves concerned with change, for example, in the theory of evolution.

I will begin with a very brief sketch of the value of Marxism and Marxist ideas in mathematics.

The development of mathematics should be studied in the first place in relation to the economic system in which it developed, and particularly to the economic needs of the society in question, and secondly, in its internal development, that is to say, in relation to the past history of methods.

Now the first of these fields has been covered to a very considerable extent by Professor Hogben's *Mathematics for the Million*. This is an excellent book in many ways; but except for a few details, it only deals with the developments of mathematics up to the early eighteenth century. It deals with them from the point of view of historical materialism, and I will only attempt to sketch some of the important lines of thought which can be followed up in that connection. Hogben stresses the importance of astronomy for people whose lives turn on some annual event, for example, the lambing season or the Nile flood. It becomes very important to be able to predict this annual event as accurately as possible. This necessitates a calendar. A calendar immediately involves you in a lot of complicated arithmetic, in attempts to express irrational numbers, for example, the number of days in the solar year, either as integers such as 365, or as simple fractional numbers, such as 365¼. Those attempts have occupied the whole working life-times of many very intelligent men. They have given rise to a lot of mystical arithmetic, such as you may find at the beginning of the Prayer Book, and other calculations of that kind. They undoubtedly led to an interest in the properties of numbers which would hardly have been arrived at by other means.

Again, it is trivial to point out that land surveying was the origin of geometry; while the further development of astronomy led to trigonometry. The very simple machines which existed at the end of the Middle Ages were yet complicated enough to arouse interest in statics; and the particle dynamics, which reached its greatest glory in the work of Newton, was very largely suggested by the study of ballistics in connection with the development of artillery.

An even more striking case arises, perhaps, in connection with another mathematical development of the late seventeenth century. At that time the theory of probability was taken up, both in England and in France. In England, the earliest work on probability was suggested by the practice of life insurance and other commercial activities involving statistics. In France, the contemporary mathematicians who worked on probability started with problems arising from games of chance played with dice or cards. The difference is clearly due to the different development of society in the two countries. In France the dominant class was still a nobility who spent a great deal of time in gambling, whereas in England the persons who were of greatest importance at the time that the Royal Society was founded were members of the rising bourgeoisie which had gained a victory in the civil war. For the subsequent development of mathematics, the problems set by the gamblers proved to be of more importance than those set by the more virtuous gentlemen who founded the art of insurance.

Nevertheless, when we have said all this, a good deal remains unexplained. With regard to the details of Greek mathematics, we can elucidate very little on these lines. We can point out that a characteristic of Greek mathematics is its abstraction, its remoteness from practical activity. To a large extent this was conditioned by the fact that in ancient Greece, economy rested on a slave basis, and most forms of practical activity were thought beneath the dignity of an intellectual. But we cannot explain the details of its development on those lines; nor can we explain a great deal of what went on in the eighteenth and nineteenth centuries when, largely as a result of the good advertisement given to mathematics by its early application to practice, particularly to navigation, chairs of mathematics were endowed, and mathematical research went on under its own momentum. It went on very much faster than any application could be found for it. There are some persons who regard such facts as this as a disproval of the doctrine of historical materialism. It is, therefore, worth reading in some detail what Engels had to say regarding the limitations of that doctrine. Here is a quotation of a letter* from Engels to Bloch, written in London in December 1890:

"According to the materialist conception of history the determining element in history is ultimately production and reproduction in real life. More than this neither Marx nor I have ever asserted. If, therefore, somebody twists this into the statement that the economic element is the only determining one, he transforms it into a meaningless, abstract, and absurd phrase. The economic situation is the basis, but the various elements of the superstructure—political forms of the class struggle and its consequences, constitutions established by the victorious class after a successful battle, etc.—forms of law—and then even the reflexes of all these actual struggles in the brains of the combatants: political, legal, philosophical theories, religious ideas and their further development into systems of dogma—also exercise their influence upon the course of the historical struggles and in many cases preponderate in determining their form. There is an interaction of all these elements, in which, amid all the endless host of accidents (i.e. of things

* *Letters of Marx and Engels* (Lawrence and Wishart), p. 475.

and events whose inner connection is so remote or so impossible to prove that we regard it as absent and can neglect it), the economic movement finally asserts itself as necessary. Otherwise the application of the theory to any period of history one chose would be easier than the solution of a simple equation of the first degree."

Later on in the same letter, he writes:

"Marx and I are ourselves partly to blame for the fact that younger writers sometimes lay more stress on the economic side than is due to it. We had to emphasize this main principle in opposition to our adversaries, who denied it, and we had not always the time, the place or the opportunity to allow the other elements involved in the interaction to come into their rights. But when it was a case of presenting a section of history, that is, of practical application, the thing was different, and there no error was possible. Unfortunately, however, it happens only too often that people think they have fully understood a theory and can apply it without more ado from the moment they have mastered its main principles, and those even not always correctly. And I cannot exempt many of the more recent 'Marxists' from this reproach, for the most wonderful rubbish has been produced from this quarter too."

I think, then, that one may be quite a sound Marxist and yet say that there are certain occasions when an analysis of the economic type will not yield anything like a complete explanation of the facts discussed. It is interesting that today mathematical practice is beginning to catch up theory to a considerable extent. If you compare modern mathematics with that of thirty years ago, you will find a somewhat larger proportion of papers devoted to fairly immediate practical problems. For example, you will notice that a very considerable amount of work is being done on problems of hydrodynamics and aerodynamics suggested by the designing of aeroplanes. You will find mathematics being applied to problems of harmonic analysis arising in telephone practice, and to statistical problems arising in the sampling of mass-produced articles.

You will also find that practice is influencing theory in new ways which are of some interest. Just because mathematics is the least practical of the sciences it has become riddled with respectability. Since the time of the Greeks, the ruler and compasses have been respectable. Anyone could use them. Other devices needed a degree of manual skill which was unworthy of a philosopher. And today certain operations in mathematics are profoundly respectable, particularly, of course, addition, subtraction, multiplication, and division. We may go farther and say that addition and multiplication are so respectable that we may even carry them out an unlimited number of times. It is respectable to give the solution of a problem in the form of an infinite series or product.

There are other operations which we can nowadays repeat more rapidly than the Greeks could multiply. Yet those operations have not got the air of respectability which attaches to addition. Let me give you a simple example. Supposing we take this cubic equation:

$$x^3 - 3x + 1 = 0$$

I was not taught at school that there is any rapid and satisfactory method of solving such equations. But supposing we write that equation:

$$x = \frac{1 + x^3}{3} = \mathrm{F}(x)$$

Then if $x_2 = F(x_1)$, $x_3 = F(x_2)$, and so on, and if x_n has a limit when n becomes infinite, that limit is clearly a root of the equation. If you can repeat that operation a large number of times, and if when you do so you get a steady result, in other words, if the iteration converges, then you have got a root of the equation.

If we put $x_1 = 0$, we can without the faintest difficulty calculate one of the roots, the successive steps being 0, $\cdot\dot{3}$, ·346, ·3471, ·3473, ·34730. That is correct to five places of decimals, and it did not take very long to do it. But you are forbidden to give $F^{\infty}(0)$ as a solution of the equation, although an infinite series is regarded as a legitimate solution of many problems. Now that tradition as to what solutions are and are not respectable is beginning to break down in a most interesting manner. One can perform simple iterations like the above almost as easily on paper as with a calculating machine, but there are certain mathematical operations which in practice can only be done by means of a special machine, and cannot be done by any ordinary person in a lifetime without such a machine. It would take years to do what the machine will do in a few minutes. In particular, Bush has produced a machine for solving differential equations of a fairly complicated character which are beyond the practical capacity of any mathematician using respectable—that is to say, non-mechanical—methods. That machine is being used to solve a great variety of problems. Among the most interesting developments of modern mathematics are Hartree's papers* in which this machine is being applied to solve problems in connection with regulation; for example, problems of how best to regulate

* *Proc. Roy. Soc. A.* 1937.

the temperature of a water bath, a furnace, or a refrigerator, or how best to steer an aeroplane blind. These problems are actually raised by industrial practice. I may say at once that Hartree's work involves a lot of complicated (and therefore respectable) mathematics, but a stage is reached in the process where this particular machine becomes necessary. That, I think, is a development of considerable interest. It suggests that possibly we are at the beginning of a new epoch of mathematics, based on a much more extensive use of practical methods than is yet considered respectable in most universities. So much for the historical side of mathematics.

With regard to the dialectical development, it can be summed up fairly simply. You discover a rule in mathematics. You next proceed to break the rule, and you then modify your original definitions in such a way as to make the breach legitimate.

Mathematics start with real, positive, scalar integers—in fact, ordinary whole numbers, such as are used in counting sheep or soldiers. Who ever heard of anybody whose lucky number was not a positive integer, but a fraction, or still worse, an irrational number such as π or $\sqrt{2}$? At the primitive level of thought exemplified by lucky numbers men started out with a primitive type of mathematics. They then learned to do certain simple operations like division.

Sometimes the operation worked. Nine goes 4 times into 36. But if you ask a child how many times 9 goes into 5, you will promptly be told that "it won't go." It therefore became necessary to invent a number which expresses $5 \div 9$ as 4 expresses $36 \div 9$. This number is, of course, $\frac{5}{9}$, or $0{\cdot}\dot{5}$. But the invention was not a simple process. The ancient Egyptians only recognized fractions with numerator 1, such as $\frac{1}{4}$ or $\frac{1}{5}$. An exception was made in favour of $\frac{2}{3}$. Thus in ancient Egypt the answer to $5 \div 9$ was not $\frac{5}{9}$, but $\frac{1}{2} + \frac{1}{18}$. And after decimals were invented, the use of recurring decimals, which are, of course, infinite series, involved yet a further step.

The early Greek mathematics were terribly embarrassed and slightly shocked by what we now call irrational numbers. We take a very simple problem such as finding the square root of two, and find that it "won't come out." The square root of two is, we say, incommensurable. It took a very long time before irrational numbers were raised to really respectable status. In fact, this was only done in the nineteenth century by Cantor and Dedekind. Again, an enormous advance was made when the number zero was raised to the rank of other numbers, and an even greater advance when negative numbers were allowed. You must remember that the introduction both of zero and of negative numbers involved the breach of rules which had been universally valid up to then, for example, that the sum of two numbers is greater than either of them, which is obviously not so when one of these numbers is zero or a negative number. It was only by revising these general rules that the new ones could be brought in. Of course, many further extensions of the idea of number have been made, such as complex numbers involving the square root of minus one, vectors with a direction, quaternions, Galois imaginaries, and many more.

The important point is that all those steps forward involved a negation of some existing rule, and the body of rules was then recast so as to negate that negation.

Just as numbers developed, so did operations. The

biggest step was possibly made by Descartes with the use of variable magnitudes. The idea is so familiar to us that we do not realize how paradoxical it seemed to Descartes' contemporaries. It was bad enough having a letter x which was going to mean some definite number when the problem was solved; but when it came to a number x which was never defined at all, it was apparent to many people that Descartes was talking nonsense. Nevertheless, that idea of a variable is one of the fundamental ideas in mathematics.

In the same way, the differential calculus involved self-contradictory notions, for example, infinitesimal numbers. In the nineteenth century, two hundred years after the invention of the calculus, fairly rigorous proofs of the elementary theorems discovered by Newton and Leibniz were finally produced. Meanwhile, other workers had gone ahead and broken more rules.

One of the most striking cases was that of Heaviside, who produced the operational calculus, in which the symbols for functions are bandied about as if they were mere multipliers. It proved of very great value in long-distance telegraphy and telephony. Heaviside could not, or did not trouble to prove the validity of his methods, although they worked; and his contemporaries in the nineteenth century, the academic mathematicians, would have nothing to do with them. It was partly as a result of this neglect that Heaviside died a very disappointed man—almost a madman. Only in the twentieth century was it found possible to render Heaviside's methods entirely respectable. This was done at Cambridge, and we now no longer learn the rule of thumb and empirical methods by which he arrived at his results.

At the moment, Dirac is employing in his work on the quantum theory a thing which he calls a delta function, which excites a certain amount of horror among respectable mathematicians, but which undoubtedly does the work for which it was designed.

Now Engels insisted on the contradictions in mathematics, particularly in the calculus. It is important to remember that he was historically correct, even although it is possible nowadays to teach the calculus in a fairly rigorous manner, as, for example, from such a book as Hardy's *Pure Mathematics*. It is, however, worth pointing out that even the so-called rigorous proofs ultimately rest on a logical basis which is not entirely free from contradiction. Students of logic learn the existence of such unfortunate facts as Russell's paradox about classes, Burali-Forti's antinomy, and so on. The rigour of modern mathematics is not, after all, as absolute as some mathematicians would like it to be.

From mathematics I pass on to physics. From the point of view of Marxist philosophy, space and time are regarded as real, and Lenin* wrote:

"Human conceptions of space and time are relative, but these relative conceptions serve as constituents of absolute truth.† These relative conceptions, in their development, move towards absolute truth. The mutability of human ideas in regard to space and time no more refutes the objective reality of either than the mutability of scientific knowledge concerning the structure and forms of matter in motion refutes the objective reality of the outer world."

* M.E., p. 143.

† I have amended the original English translation, which is far from accurate.

However, it is worth adding that, according to Engels:*

"There must be things which have shape and whose shapes are compared before anyone can arrive at the idea of form. Pure mathematics deals with the space forms and quantity relations of the real world; that is with material which is very real indeed. The fact that this material appears in an extremely abstract form can only superficially conceal its origin in the external world. But in order to make it possible to investigate these forms and relations in their pure state, it is necessary to abstract them entirely from their content, to put the content aside as irrelevant; hence we get the point without dimensions, lines without breadth and thickness, a and b and x and y, constants and variables; and only at the very end of all these do we reach for the first time the free creations and imaginations of the mind, that is to say, imaginary magnitudes. Even the apparent derivation of mathematical magnitudes from each other does not prove their *a priori* origin, but only their rational interconnection. Before it was possible to arrive at the idea of deducing the form of a cylinder from the rotation of a rectangle about one of its sides, a number of real rectangles and cylinders, in however imperfect a form, must have been examined. Like all other sciences, mathematics arose out of the needs of men; from the measurement of land and of the content of vessels, from the computation of time and mechanics. But, as in every department of thought, at a certain stage of development the laws abstracted from the real world became divorced from the real world, and are set over against it as something independent, as laws coming from outside, to which the world has to conform. This took place in society and in the State, and in this way, and not otherwise, pure mathematics is subsequently applied to the world, although it is borrowed from this same world and only represents one section of its forms of interconnection; and it is only just precisely because of this that it can be applied at all."

* A.D., p. 47.

How, we may ask, does the principle of relativity fit in with those ideas? There is no question that idealist interpretations of that principle have been given. I have even given one myself in the past. Space and time, say the idealists, are shown to be human constructions. Einstein has shown, it is claimed, that different observers "have" different spaces, and on the basis of that, Eddington in particular has developed the ideas of Kant to cover a very wide field. We must certainly allow Eddington a measure of truth. Even if one thinks that he is standing on his head, in that position he can perform intellectual operations which the rest of us cannot perform standing on our feet! Further, I think it is clear that certain elements in our usual account of space and time are human constructions. However, it seems to me that these elements are not given in perception, but occur in the somewhat abstract systems which geometers have built up. So after all, perhaps Eddington will turn out to have been throwing out the baby with the bath water.

From a materialist point of view, it seems to me that relativity is quite intelligible, once we regard the world as consisting, not of things, but of processes or events. We then observe in many cases that an event A occurred before an event B and influenced it. In such cases the principle of relativity never reverses the order of those events, or makes them simultaneous. The doubt arises in other cases. Supposing that our two events A and B occurred on different stars. A is the explosion leading to the production of a nova (a new star) in a spiral nebula in one direction, say the Great Bear, and B is a similar event in a spiral nebula in an opposite direction, say the Southern Cross. These events may be so far off in space and so near in time that light started from A will arrive at the star where B occurred long after B, and also the light starting from B will arrive at the site of A long after A. As we have strong reason to believe that nothing can travel more rapidly than light, no direct causal relation between the events A and B can be possible. Neither can influence the other. They may influence events in their common future. Now the principle of relativity states that some observers will regard A as having happened before B, and some will regard B as having happened before A. But from the point of view of the theory of relativity, no absolute truth can be attributed to either of those statements. It is important to realize that this denial makes no practical difference, because there is no causal connection between events A and B. Similarly observers in motion relative to one another will use different measures for spatial intervals in the direction of their relative motion.

The theory of relativity actually arose through the necessity of making theory conform to practice in certain cases. It became clear that two observers in motion relative to one another, for example, one in England and one in New Zealand, both using Euclidian space and Newtonian time, would differ as to what events they considered simultaneous; and there is no way to decide which was right, because there is no criterion of absolute rest. The Newtonian theory of relativity was, of course, familiar. The fact was well known that it made no observable difference to the internal mechanical behaviour of a system if that system was in uniform motion, and that there is therefore no way of determining from mechanical observations what systems are at rest. Einstein's theory was an extension of Newtonian relativity; and it turned out that simultaneity, as well as rest, was an idea to which no practical meaning could be attached. In other words, it followed that the classical theory of space and time must be rejected as being metaphysical, that is to say, as postulating something beyond matter, namely an abstract space and time which had properties apart from those of any events going on in them. Such a postulate could only be justified if we knew of the existence of something more fundamental than matter, using matter in the broadest sense to cover processes as well as things.

Professor Cook-Wilson, who attempted to teach me philosophy, believed that he had absolute *a priori* knowledge of space, and on the basis of that *a priori* knowledge he was able to refute Einstein to his own complete satisfaction. Nevertheless, I doubt whether, in spite of certain statements to the contrary, a believer in space and time existing apart from matter would be a good Marxist. I do not think that a Marxist need object to a very considerable modification of our ideas of space and time. On page 56 I quoted Lenin's views on space and time, written a year or two before the publication of Einstein's first theory of relativity. I have been unable to obtain a translation of an article which he wrote in 1922, in which I understand that he accepted relativity, but rejected idealistic interpretations of it.

"Lenin's opinions on this matter are only one example of his amazing capacity for sticking to essentials while discarding theories or practices which seemed fundamental to people of less penetration. In the sphere of practice this is illustrated by his New Economic Policy of 1921, which was criticized as a surrender of socialism and a restoration of capitalism."

As a matter of historical fact it led to the complete victory of socialism in the First Five-Year Plan. In the same way a retreat from some of the untenable positions of Newtonian mechanistic materialism is a prerequisite for an advance to a still more thoroughgoing materialism.

It is only fair to add that it would very likely be possible to quote isolated sentences, both from Lenin and Engels, which would hardly be compatible with a full acceptance of the theory of relativity, though I have not myself discovered any such passages. And that would, I think, be a fatal objection to Marxism as a contribution to scientific method if it was not a part of Marxist theory that Marxism lays no claim whatever to finality.

We may take it, I think, that Einstein's greatest contribution to scientific pnilosophy is the simple principle that if a thing or a process is essentially unobservable, it does not exist. By the application of that principle, he cleared away a good deal of what was thought to be essential in our knowledge of nature, but which turned out, if he is correct, to be merely intellectual scaffolding. As we shall see, however, we may yet be able to attach a meaning to simultaneity by a fearless and dialectical application of Einstein's methods. On the other hand, he left, in the opinion of most physicists, what was really essential, although there are thinkers such as Eddington who go a great deal farther than Einstein, and reject what the majority of scientists regard as fundamental truths. It is worth noting the primacy of practice over theory in Einstein's work, and his essentially anti-metaphysical bent, that is to say, his refusal to take seriously things like absolute space and simultaneity which cannot be observed. The end result is a considerable simplification. The special theory of relativity is certainly simpler than the theory that material objects contract in the direction of motion when they move. But since Einstein's special theory, there have been many further developments along the same line by Einstein and others which, if confirmed, will give us positive additions to knowledge.

I propose to describe briefly one of those theories, the most interesting, certainly the most ambitious, of them, namely the theory of Milne, not with any suggestion that it is final, or that it will not be found to contain a number of errors, but in order to give an idea of how cosmological theory is actually growing at the present moment. Milne starts from the idea of what he calls cosmological relativity. That is to say, the idea that the universe would appear similar to observers on a number of different heavenly bodies and particularly on different spiral nebulae (the large groups of stars of which our own Milky Way is one containing, among millions of other stars, the sun and the planets circling round it). If cosmological relativity is true, it means that nowhere, however far one went in space, would matter come to an end, nor does it progressively thin out in one direction. In other words, we observers on this planet are in no way privileged. It may be regarded as an extension of the disproof of idealism which appealed most strongly to Lenin, a disproof on the ground that it leads to solipsism, the theory that you are all in my dream, a theory which even if true is certainly discourteous! Milne tries to apply that standpoint to the universe as a whole, taking the view that there are no privileged observers who are in the middle, while others are at the edge. It follows at once that unless space is finite there is an infinite number of observers, and therefore of spiral nebulae.

His theory is almost certainly not strictly true, if only because it becomes clearer every day that the spiral nebulae are not distributed entirely at random. Nevertheless, this principle may be regarded as a very good approximation.

He starts with this simple postulate, and imagines observers on different spirals who can signal to one another with light. This would be rather expensive, as we can see no event less colossal than the expansion of a star, even in the nearest of the spirals. He assumes that the velocity of light is constant, an assumption which merely serves to connect his time and space scales in a simple way, and which agrees well with ordinary practice. He further assumes that space is Euclidian.

Now if all the observers were uniformly distributed through space, and at rest with regard to one another, the universe would be blotted out in a blaze of radiation. However, the principle of cosmological relativity can be satisfied if they are all packed, according to a particular law, into a sphere which is either expanding or contracting with the velocity of light. In the case of the expanding sphere, which agrees with observation, all the observers are moving away from one another. Each one appears to be at the centre of the sphere, and is so provided he or she uses the system of measurement which he would naturally use for objects in his immediate neighbourhood. This curious result follows at once from the special principle of relativity. Thus there are no privileged observers, though each one has a good excuse for thinking himself privileged.

When we observe the spiral nebulae we find that they seem to be composed of stars much like our sun and its neighbours, and their spectra show the presence of the same chemical elements. But the spectral lines are all shifted towards the red, as are the spectra of stars which are known to be moving away from us. The speed of recession can easily be calculated from the shift, and it is found that the few bright nebulae which are believed to be nearest, are moving away most slowly, while the many dim ones which are thought to be farther off, are receding more quickly. If the speeds are uniform, then about two thousand million years ago all the nebulae would have been concentrated into a small volume.

Several other cosmologies, including Eddington's, are in agreement with this view, but they do not agree with Milne's on other points. According to Milne, on the scale of measurement appropriate to any particular observer, space would appear infinitely overcrowded near the expanding edge of the material universe. In the last millimetre there are an infinite number of nebulae, flattened out in the direction of their motion, and there is no farthest one in any direction, any more than, for example, there is a largest number. This apparent over-crowding is, however, only a paradoxical effect of the system of measurement. To a local observer moving with the appropriate speed everything would seem normal.

Going back two thousand million years into the past, all the infinite number of atoms in the expanding universe were packed into an indefinitely small volume, from which they rushed out explosively. This explosion is identified by apologists of religion with the creation of the world by God, though it is not very like the accounts of that event given in the Bible and other ancient documents. The same figure of two thousand million years turns up in many other connections. The atoms of the metal uranium break down at a constant rate, yielding lead and helium as end products. Analyses of rocks show nothing older than about fifteen hundred million years. The helium in meteorites agrees with an upper limit of two thousand million years for their age. Calculations regarding the orbit of Mars give much the same age for it.

Milne was satisfied with this cosmology in *Relativity, Gravitation and World-Structure,* published in 1936. It enabled him to deduce from simple postulates the existence of a great variety of phenomena, from gravitation to cosmic rays. And as he is a theist, the idea of creation did not trouble him. Among the oddest features of his cosmology is the gradual diminution of the gravitational constant, and the increase of angular momentum. Thus on this hypothesis concerning time, the year and the day would progressively lengthen, and the earth recede from the sun.

But Milne is an intellectually honest man. And in a series of papers recently published in the Proceedings of the Royal Society he has followed up the consequences of his own thought, and shown that most of these oddities, and particularly the overcrowding at the time of "creation" and near the expanding edge of the universe, can be made to disappear completely. The expanding universe theory became slightly ridiculous when he deduced that owing to a diminution in electrical forces, atoms were expanding at just the same rate as the rest of the universe, i.e. had doubled their radius in the last thousand million years. In other words, if we take an ordinary object such as a standard yard as our scale of length, the universe is not expanding.

In order that ordinary objects should not increase in size we must alter our time scale. Instead of using the kinematical time scale he uses a dynamical time scale in which the time is measured by $\tau = t_0\left(\log \frac{t}{t_0} + 1\right)$, where t_0 is the present time on the kinematical scale, i.e. 2×10^9 years. At the present time the two scales are equivalent. It does not matter which we use to measure the year. But while the kinematical time scale is finite in the past, dynamical time is infinite in both directions. This is an elementary property of logarithmic scales. Thus if we arrange musical tones according to their frequency, the scale comes to an end at a frequency of zero. But if we use the piano scale (more accurately the chromatic scale) where the distance between two tones is proportionate not to the difference of their frequencies but to the difference of the logarithms of the frequencies, we find that it goes on for ever in both directions.

It is worth noting that perception gives us the order of temporal events, but does not give us an accurate time scale. Subjective time depends on the rate of a particular chemical process in our brains. A man with natural or artificial fever overestimates the number of seconds in a given interval. Indeed the brain temperature can be determined rather accurately by making a person estimate a minute. The heart also speeds up, but according to a different law. There is therefore nothing sacred, so to say, about any particular time scale.

To return to Milne's cosmology, if we adopt the dynamical time scale we find that atoms are not expanding, nor is the universe. The length of the day and year remains constant save for the effects of friction. The spiral nebulae are not flying apart. And there was no creation at any time in the past. Time stretches backwards and forwards for ever. But two important modifications of our ordinary physics must be made.

Space is no longer Euclidian, but hyperbolic. Euclid's postulate about parallel lines is untrue. The sum of the angles of a triangle is very slightly less than two right angles, and the area of a circle is increased more than fourfold when its radius is doubled. This geometry was worked out by Lobachevski a century ago, and is known to be just as consistent with experience as Euclid's. The deviations from Euclidian geometry only become appreciable at distances measured in thousands of light-years, just as deviations from the flat-earth theory do not become appreciable if you are making a map of Birmingham, but are quite important for a map of England.

It is important to note that it makes no difference at all to our predictions of observable phenomena which of these two schemes we adopt. Such indifference is often attacked by materialists as a concession to positivism. I believe that this attack is unjustified. Space is real as the system of relationships between material objects or events. But it has no absolute existence apart from matter, and a belief in its existence apart from matter is a step away from materialism towards metaphysics. So with time. The order of events in time within a given material system is an objective fact. The scale on which they are to be measured is a matter of convenience. This would not be so if we could move a piece of matter, say a tape measure, from one place to another without altering it, or if we could be sure that an event, say the swing of a pendulum in 1938, was the same as its swing in 1930. But neither of these is true. The one idea is undialectical, the other unhistorical.

But if we say that it is a property of space that a finite volume of it does not contain an infinite number of atoms, and of time that a finite interval does not contain an infinite number of events such as pendulum swings, we must certainly choose dynamical time and hyperbolic space rather than kinematic time and Euclidian space. Of course, still other modes of measurement are possible.

The most remarkable feature of Milne's cosmology is that whichever set of scales we adopt, it is clear that the laws of nature are changing. If we adopt the kinematical scale, we find that the gravitational constant is altering. If we adopt the logarithmic or dynamical time scale, we find that the distant nebulae are redder than the near ones not because they are running away, but because atomic processes are speeding up. Old light is red because though dynamical events such as the earth's rotation are proceeding uniformly, atoms were vibrating more slowly ten million years ago than nowadays. If, then, dynamical changes are to proceed at a steady rate, with days and years constant, chemical changes and radio-activity must be speeding up, and if we go far enough back, we find that matter, though many of its physical properties were as they are now, was chemically inert.

There was a day when atoms were really atomic, and refused, except very rarely, to be broken up into electrons and nuclei, as they can be nowadays. Or if electrons were free from the atomic nuclei, they only united with them on very rare occasions. If that is so, some rather remarkable results follow. They also follow from a theory which is being developed by Dirac on a very different basis. Now Engels, in a letter to Marx, dated May 30, 1873, suggested the following dialectical chain. Starting with the motion of separate bodies, under gravitation, like that of the planets round the sun, he pointed out that when they came into contact, the original laws were suddenly negated, and new laws of a more complicated character came into play, and that thirdly, as a result of that contact, electrical and chemical changes took place which would not have occurred had the bodies not been in contact. Therefore, from his point of view, chemistry is newer than physics in world history as well as in human thought and practice.

Engels did not, like Hegel, have the audacity to try to represent this as an actual picture of world development. He realized that when Hegel tried to apply his philosophy in too great detail, he made some very big mistakes, and in consequence he did not develop that idea any farther. Indeed Engels toyed with the idea which was common during the nineteenth century that the various parts of the world went through phases of development and dissolution, that there was in fact no history of the universe as a whole. It is worth pointing out that he never published this speculation, which occurs in the notes for *Dialectics and Nature*. Obviously he was not satisfied with his results.

It would now seem as if we really could point to an historical process which applies to the universe as a whole, as if, in fact, the laws of nature themselves were gradually changing. If so, it would follow, to take an example, that the radio-activity of any long-lived element passes through a maximum and then fades away, constantly diminishing. It would follow that it is only since a finite period in the past that the chemical properties of matter have been of such a character as to permit life. If that is so, a number of very remarkable facts collected by L. J. Henderson in his book *The Fitness of the Environment* would find their explanation. Henderson pointed out that a number of the properties of matter were particularly adapted for life, and that existing scientific theories could only suggest that these were coincidences. If, however, this collocation of properties only occurs at a certain stage in the history of the universe, we have to deal with a fact like the obviously provable fact that large towns are often found on rivers and rarely on mountain tops. Life is at present possible because matter has its present properties. We have to take an historical view of the properties of matter and the so-called laws of nature. In the future we should expect to find the development of new properties in matter, properties which at the present moment could only be determined by the most refined methods, but which matter will exhibit in a developed form a few thousand million years hence. In fact, the universe would appear to have many more surprises in store, not for us, but for our distant successors, than appeared possible from the point of view of scientists who believed that the laws of nature were eternally fixed.

In classical physics, including ordinary relativity theory and quantum theory, time occurs in the equations in such a manner that it makes no difference whether an event occurs now or a hundred million years hence. There is no historical element in physics, as ordinarily taught, whereas there certainly is in biology. It makes a considerable difference to a dog's behaviour when you hit him, whether it is the first time or the hundredth time that you have done so. If it is the hundredth time, he will almost certainly dodge. Apparently although it makes a difference to a dog, it makes no difference to a kettle of water whether it is the first time or the hundredth time that you boil it. There are a lot of physical and chemical processes which appear to be reversible, like the boiling of water. But if Milne's cosmology is accepted, reversibility, although an absolutely satisfactory postulate for ordinary purposes, is in the long run inapplicable.

I think it is possible that whatever may be the final verdict on Milne and Dirac's work, it will be found that they have introduced the historical process into exact physics, even although for all ordinary purposes these historical changes will be negligible.

It further follows that the idea of simultaneity can be given a meaning once more if events are said to be simultaneous when they take place in systems obeying identical sets of natural laws. Milne, of course, realizes that he has brought back the idea of "cosmical time." He probably does not realize the beautifully dialectical nature of the argument. Einstein negated the notion of simultaneity. He pointed out that there was no way of testing whether two events are simultaneous. By a thoroughgoing extension of Einstein's principles Milne has made it probable that a way of carrying out this test may exist. If this is true Einstein's negation has been negated.

I must add that neither Milne's cosmological theories nor those of Dirac have been generally accepted by physicists. This is entirely as it should be. They must be tested by their agreement with observed facts and above all by their capacity for predicting facts hitherto unobserved. Nevertheless it is difficult to believe that some at least of their conclusions will not be incorporated into the general body of scientific theory.

Many other developments of modern cosmology are in close harmony with Engels' thought. The so-called fixed stars were first classified empirically by their spectra. In some of them lines due to helium predominated, in others hydrogen was more obvious, in yet others carbon or manganese. It looked as if the stars could be divided into classes of a different chemical composition, like minerals. But Saha and others pointed out that as the temperature of a star's photosphere varied, atoms of different elements would be in such a state as to absorb light and leave their traces in the spectrum. The different stellar types were taken to be different stages in the evolution of a star. The stars were supposed to start as diffuse and relatively cool spheres of gas, and to heat up as they contracted, getting bluer as they did so. After passing through a maximum temperature they cooled down and got redder again. This theory of continuous evolution was perfected by Hertzsprung and Russell.

But here, too, Milne has complicated matters. Occasionally a star explodes, the explosion being described as a nova, or new star, which sinks back to near its original brightness in a few months. Milne thinks that this event may be expected to occur once in the "life" of every star which lasts long enough. A star of about the size of our sun suddenly collapses into a "white dwarf" of extreme density, and with a considerably hotter surface. The energy due to the collapse is given out in radiation. The development of a star thus includes a good deal of steady evolution and at least one revolutionary "leap." It is still too early to say how far this theory will have to be modified in the light of the new conceptions of time. But at least it is clear that the stars must be viewed as historical processes, not given facts.

3. *Quantum Theory and Chemistry*

I MUST now pass on to an entirely different branch of growing physics, namely quantum theory. Few branches have suffered worse in presentation than the quantum theory, particularly because for practical purposes mathematical physicists who are working at it use mathematics up to the very last moment when they see whether or not their mathematical predictions are verified; and when they try to put these mathematics into words they sometimes talk nonsense. But in addition, there has been an undue tendency in certain popular books to exploit its paradoxical and negative sides, whereas its positive side gives information additional to that of classical physics, which is just as real.

In order to understand even what the quantum theory is about, we must study the historical development of physics. In the eighteenth century physicists held that the world consisted of particles, some of which were ponderable, while others, such as those of light, were imponderable. There might also be imponderable fluids, such as heat. Further, it was thought in the eighteenth century that these particles, like ordinary visible particles, would move in straight lines unless acted on by some force, and that the force depended on the position of other particles. The most complete exposition of this way of looking at things was Laplace's *System of the World*. Laplace said that if a super-human calculator knew the position and the speeds of all particles in the universe, he could then predict the future exactly. Laplace, who claimed that he had banished God from his system of nature, brought Him back in this ghostly form to establish determinism. The deists, who had no use for God for any other purpose, kept Him to start the world machine.

It is worth pointing out that neither the God of the deists nor the calculator of Laplace can possibly be equated with the God of ordinary religious people, who at least serves to prevent the world being a mere machine. Now this mechanical view has been pretty generally accepted by scientists since Laplace's time; and even those who did not believe in the existence of the "high all-seer," to use Shakespeare's phrase, thought that the more we knew about any material system, the more accurately we could predict about it, and further that there was no limit to the possible accuracy of our predictions. This principle is very obviously nearly true in many cases, notably for ordinary astronomical or chemical phenomena. It would, however, be extremely rash to apply it all through the sciences.

In the nineteenth century, the unity of the eighteenth century particle physics was broken by the wave theory of light. The wave-like properties of light are demonstrated by a phenomenon of interference, such a phenomenon as you can observe by looking at a distant light through muslin, or by experiments with a grating ruled with fine parallel lines. This preferentially reflects light of any particular colour when this light strikes it at such an angle that the successively reflected waves reinforce one another. It was awkward when, towards the end of the nineteenth century, it was found that light had mass, exercising a push on a surface which was reflecting it. This would have been expected if light were a stream of particles, but could also be explained on the wave theory. Early in the twentieth century, however, the fearful discovery was made that light, although it had these wave-like properties, also had properties similar to a stream of particles. It was found that when light was shone on certain metals like potassium, electrons jumped out of those metals. This is the principle on which the ordinary photo-electric cell is based.

It was further found that the speed with which the electrons were shot out was constant if the light was monochromatic, that is to say, of a single colour, and was proportional to the frequency of the light. On the other hand, the number of particles shot out was proportional to the intensity of the light. The light behaved as if it consisted of a stream of particles whose energy E was proportional to the frequency ν ($E = h\nu$). I may add that this equation is generally known as Einstein's equation, so it is clear that not every formula due to Einstein is terribly complicated or unintelligible. The constant *h* (Planck's constant) is constantly turning up in connection with the transition between the wave-like properties and particle-like properties of things.

It was later shown that electrons behaved in certain respects like a system of waves whose frequency obeys the same law, although there was no doubt about the particulate character of the electrons. You can get one, two, or three into a drop of oil and measure its charge accur ately enough. Intermediate charges are never found.

From these facts developed the Uncertainty Principle, which is the one principle in quantum theory that has been rather heavily popularized. It may be summarized thus. Suppose that some very fine particles, like *Lycopodium* spores, are being shot out through a small hole into a vacuum. You want to measure their speed. The obvious way is to take two photographs at a known interval—say one-thousandth of a second, by flashes of light, and then find out where a particle was on these two different occasions, and how far it has gone during the interval. Now if one uses red light under the microscope one gets a fuzzy image. It will come out a good deal clearer if one uses ultra-violet light. So one would need, in such an experiment, in order to be as accurate as possible, to use light of a small wave-length. But unfortunately it is a known fact that small particles, for example, *Lycopodium* spores, are deflected by radiation. In fact they are pushed away by a beam of light hitting them. The obvious remedy for that is to cut down the intensity of the light we are using for our photograph. We can do that up to a point, but unfortunately we cannot expect to get any image on our photographic screen unless at least one packet of light has been stopped by the thing which we are photographing. Everything would be all right if light had not got this horrible habit of interacting with matter in packets of a definite size, given by Einstein's equation. So we have the awkward dilemma that the sharper our image, the bigger the bump the particle is bound to get when we photograph it. Therefore, the more accurately we take our photograph, the more the speed of the particle must have been altered when the first photograph was taken.

Let us say at once that for a particle of the size of a spore, this principle does not lead to any very serious uncertainty. We can determine its speed quite accurately enough for most purposes. Perhaps, indeed, the errors introduced by the uncertainty principle are less than those which arise from quite other causes in most of our measuring instruments. But when we get down to particles as small as electrons, these difficulties become very serious indeed. The wave-like properties of ordinary matter conspire to give us exactly the same uncertainty if we decide to substitute some other method for our photographic methods, for example, the opening of a shutter in a wall.

It is worth pointing out that according to the uncertainty principle, we can measure the position of our particle on the two occasions of photographing it, and its average speed between these, as accurately as we want to. There is no essential fuzziness about matter which prevents us observing it as accurately as we wish. On the other hand, when our particle goes off after its second photograph, we do not know how much its velocity has been changed by the act of photographing it; similarly we do not know its speed before the first photograph was taken. In fact, we can tell what it was doing during the interval that it was being observed, but cannot tell what it is going to do next. In other words, our observation of any object is, among other things, a physical process which affects the object observed. It follows that there are no observers who are only observers and merely sit back and take no part in the processes of the universe. That is an extremely general principle of Marxism. Marx continually pointed out that observers of society are also active members of that society, that either they are producers or they do not produce. In each case, it will make a difference in their outlook. In general, it is impossible to get outside the universe. In fact, the existence of a "high all-seer" who does not act is self-contradictory. Mind is always active as well as passive. And perception always involves physical processes.

Now it is noteworthy that the doctrine of determinism, the doctrine that the future is certain, was originally not a scientific but a theological doctrine, arising from the theory that God knew everything. To my mind, it is meaningless unless there is an omniscient being who will impart some of his knowledge to men, because there is no way of testing it. It is perfectly true that you can ask whether a given person can predict the future with a particular degree of accuracy, whether, for example, he can say that the number of people killed on the roads of England this year will be between five and seven thousand. But the question of whether all future events are absolutely determined today seems to me meaningless unless it can be tested. And unless it can be tested by reference to the knowledge of an omniscient being I can see no method of checking it as ordinary scientific theories are checked. And, therefore, according to Einstein's principle, or, if you like, according to the principle of the unity of theory and practice, we had better scrap the idea of determinism, at least as understood by Laplace. This does not, of course, mean that a very high degree of accuracy is not possible in the prediction both of physical and social happenings.

Further, since human knowledge involves physical interaction between two material systems, it is difficult to see how an omniscient being can be fitted into the universe which we are beginning to know. In spite of the views of Jeans and Eddington it seems to me that quantum mechanics raise more difficulties for theism than they solve. However that may be, it is of the utmost importance to point out that quantum mechanics gives us a great deal of new knowledge, knowledge of a kind which seemed quite impossible before. What it takes away with one hand, it certainly gives back with the other.

According to nineteenth-century physicists, it was impossible for an electron to go on circulating round a positively charged nucleus. It would gradually give out its energy in radiation. But if it can only do that in packets whose size is determined by quantum relations, then we can begin to predict a great deal more as to what sort of orbits are possible, and by a very considerable extension of the quantum principle, which is entirely impossible to explain in less than several hours, it has become possible to calculate with great accuracy what orbits are possible in a given system of negative electrons circulating round a positive nucleus. There is, in particular, a minimum size for these orbits, and in consequence the atom has got a certain minimum size which is precisely defined. There is no reason from the point of view of the nineteenth-century physicists why a system of particles, as opposed to solid bodies in contact, should have any particular size; yet rigidity of solids is one of the most general facts of nature. It is only the quantum theory which has made this intelligible.

In the same way, we know from the study of spectra that an excited atom will only give out energy in packets of a certain size, and therefore produces sharp spectral lines. That, again, was entirely unintelligible to the nineteenth-century physicists. Yet it is now possible to predict spectra which have not been observed, and to find them. The conception of minimum energy is one of considerable importance. By no amount of cooling can we get the last trickle of energy out of matter. As Engels* put it: "Motion is the mode of existence of matter. . . . Matter without motion is as unthinkable as motion without matter."

Actually, although we certainly cannot predict the future movement of a given electron, we can predict the distribution of a number of electrons with an accuracy which is very great indeed, if the number of electrons is big enough. This principle is constantly applied by Marx and Engels to society. It is one of the bases of their social philosophy.

To sum up the quantum theory from this point of view; Engels† wrote fifty years ago: "One knows that what is maintained to be necessary is composed of sheer accidents and that the so-called accidental is the form behind which necessity hides itself." In our case the

* A.D., p. 71. † F., p. 55.

necessity is the sharp quantization of atomic energies, which leads to sharp spectral lines and almost rigid atoms. The accidents are events in individual atoms which, however, add up to practical certainty when we are averaging trillions of atoms, as is almost always the case in practice.

This same fact which gives us the uncertainty principle in connection with the motion of individual particles gives us an immense amount of verifiable deductions concerning properties of matter which add to our knowledge to such an extent that we might legitimately speak of a new certainty principle.

One more fundamentally dialectical characteristic of quantum mechanics must be noted. Supposing you have two like elementary particles, such as two electrons or two atomic nuclei. They approach one another very closely and then separate. We cannot be quite sure which is which, because there is no way of labelling an electron. Now the interesting principle of quantum mechanics concerned in such processes is that this is not a mere consequence of faulty observation, or of our inadequate means of observation. It shows itself in the actual behaviour of these particles. For example, the alpha particles which are shot out of some radio-active elements are the nuclei of helium atoms moving very fast. Now when we shoot these particles through a gas, they are scattered according to a definite law, and that law is the same for all gases except one, for hydrogen, nitrogen, oxygen, and so on. The one exception to the rule is found when you shoot them through helium, because in this case the alpha particle is not distinguishable from the nucleus of the atom which it hits. Therefore, this uncertainty is not just a human imperfection. It is something rooted in the nature of matter. We have to conclude, therefore, that the individuality of these elementary particles is not anything absolute. Along with that individuality you find its negation, which becomes important when two particles approach very closely.

Now I do not want to suggest that Engels predicted quantum mechanics. Engels had before his eyes the awful example of Hegel when he ventured into the philosophy of nature, and in consequence he steered very wide of too detailed predictions in the scientific field. Certain of the principles of quantum mechanics are not particularly dialectical, so far as I can see; for example, Pauli's exclusion principle. But the point which I wish to make is that no other philosophical theory needs so little modification in view of the facts discovered about quantum behaviour as that of Engels.

It is worth comparing the reactions to it of people holding idealist views. Sir James Jeans says: "Nature consists of waves, and . . . these are of the general character of waves of knowledge, or of absence of knowledge, in our minds." It seems to me that this is a somewhat muddled point of view, to put it mildly. That of which we are ignorant, namely, the position of electrons, is not apparently part of nature at all! It is a very significant social fact that the book which contains that quotation is a best-seller.

I should like to conclude my brief review of this question by quoting from what seems to me the most intelligible book to an ordinary reader which I know on the subject of quantum mechanics, although as it was published in 1931 and is, therefore, to some extent out of date. This is C. G. Darwin's *The New Conceptions of Matter*, which (another interesting social phenomenon) is *not* a best-seller:

"There have been several excellent books recently written on the subject, but it has seemed to me that some of them, perhaps through attempting to cover too wide a field, do not always succeed in explaining their subject. They excite the wonder of the reader, by suggesting to him what extraordinary difficulties there are in the idea of physics; they are like a conjuror whose tricks seem to us inexplicable. I have set myself what is, I think, a more ambitious task, for I want to try and show how the tricks are done. I shall count myself as having succeeded, if at the end of the book any surviving reader will speak no longer of the mysteries of science, but, shall we say, of the naturalness of Nature."[11]

I should like to suggest that when we really begin to examine quantum mechanics and to speak about its results with less bated breath than we do at present, it will turn out that after all, nature is natural.

During the nineteenth century, there was a sharp distinction between chemistry and physics. It is true that the chemists employed physical methods, for example, they carefully measured the weight, temperature, volume, and so on, of the substances with which they were dealing. They measured the current of electricity passing through a solution and the amount of metal which it deposited on the cathode, and, of course, they used the existing physical theory up to a point. But in addition, they used concepts such as that of affinity which had no place in nineteenth century physics. Comte and the positivist school which he founded insisted that there was a sharp separation between physics and chemistry, and an equally sharp gap between chemistry and biology. The modern logical positivists say that a sharp distinction should be drawn between physical language, biological language, and so on.

Engels and other Marxist writers, along with, I think, the majority of practising scientists, refused to accept that view, although they could not say a great deal as to how they expected the gulf to be bridged. One of the main reasons for the sharp separation of chemistry and physics was that the rigidity of solids compelled the physicists to regard atoms as being rigid structures. They generally thought of them as something like little billiard balls with hooks attached to them to allow of combination, and there was no place in the physical scheme for the properties which the chemists studied.

There were two ways of trying to express the relationship between physics and chemistry. One was to say that chemistry is physics plus X. In chemistry the physical laws hold, and there are also a whole set of other laws. The second line of approach was to say the concepts of physics were not yet a near enough approach to the truth to explain the facts of chemistry. If that was so, we should expect the progress of physical research to develop its own internal contradictions, until the physicists were forced, not so much by a consideration of chemistry as by the internal necessities of their own science, to state it in such a way that it became possible to explain chemistry. Of course, it was the latter that happened.

I gave a brief account on pages 74–78 of the difficulties which forced the physicists to put forward a view of matter that was capable of explaining the facts of chemistry. A great deal of elementary chemistry is explained quantitatively by means of quantum mechanics in a way which a few years ago seemed quite impossible to such a philosopher as Broad, and to others who think on positivist lines, even if they do not call themselves positivists. Broad insisted that the difference between the properties, say of sodium chloride and silver chloride, was something unpredictable from our knowledge of silver and sodium; and it is quite certain that the properties of sodium chloride and silver chloride were investigated a long time before it was found out that they could be explained. Nevertheless, it is now possible from our knowledge of the spectrum of an element, the list, that is to say, of the kinds of light which it emits when excited, to deduce a great deal about its chemical behaviour. We can deduce that sodium, potassium, and other alkali metals will readily lose one electron and therefore form univalent bases. What is much more striking, fairly exact calculations of chemical affinities and heats of reaction have been made on the basis of the quantum behaviour of the atoms as deduced from their spectra. Chemical bonds, which the ordinary chemist represents by lines between the symbols for two atoms, may either be due to the transfer of an electron from one atom to another, as when sodium transfers one to chlorine, giving two ions somewhat stabler than the uncharged atoms, and held together in the solid by electrostatic attraction. Or they may be due to a sharing of electrons either in a pair or some other number, which is almost always even. For example, in a chlorine molecule, two are shared, and confining our attention to the outer shells of electrons, we symbolize it as $:\underset{\cdot\cdot}{\overset{\cdot\cdot}{Cl}}:\underset{\cdot\cdot}{\overset{\cdot\cdot}{Cl}}:$

Now the complex behaviour of a molecule, far more

complex than it has any business to be if molecules were as simple as physicists and even chemists believed, does not depend on complexity of structure as expressed in terms of nuclei and electrons, but is fully explained by its complex energetic structure. The energy of rotation and the energy of various kinds of vibrations do not grow continuously with temperature; they are added in small steps according to quantum laws. A rotating molecule cannot have any given angular velocity, it can only have an angular velocity with certain definite values, so the energy of rotation can only be got rid of in little packets; and in that way we account for the complex spectra and a great many intricate details of the behaviour of chemical substances. In particular, it is important to realize that a number of predictions have been made on the basis of quantum mechanics, and some of them, at least, have been verified. Among the most striking is that of the two isomeric forms of hydrogen. I am not referring to the difference between light and heavy hydrogen, but to the difference between ortho- and para-hydrogen, discovered somewhat earlier. One of these can be transformed into the other, and the transformation causes an appreciable difference in the specific heat at low temperatures and in other properties. This is a fact which was absolutely unexpected by the chemists and physicists who had been studying hydrogen for more than a century, until it was predicted on the basis of quantum mechanics.

The relation of chemistry to industry is well known, though not so well understood. If I had the time and the knowledge I would particularly stress the way in which organic chemistry and chemical industry have aided one another in Germany, and the close connection between Imperial Chemical Industries and research work in Britain today, a connection which has both its good and its bad side. I should point out, in particular, how organic chemistry started as the study of chemical compounds found in living organisms, and developed into the study of synthetic carbon compounds largely under the influence of industrial demand, until the old organic chemistry was resurrected within the last thirty years under the name of biochemistry, which in turn has become to a considerable extent bound up with the food and drug industries.

In so far as dialectic is concerned, Engels' account of its application to chemistry is still well worth reading. He owed his chemical knowledge to Schorlemmer, the first communist Fellow of the Royal Society, and the collaborator with Roscoe in a well-known text-book. I shall touch briefly on some dialectical developments of chemistry since Engels' time. He stressed the immense importance of Mendeleyev's periodic law, according to which the chemical properties of elements are periodic functions of their atomic weights. But for nineteenth-century chemistry the elements were something given, eternal bricks of the universe with no history. Engels did not believe this, but he could not disprove it. We now know that the classification of the elements is not so clear-cut as was thought. The atoms with given chemical properties have nuclei with identical charges, the number of unit positive charges being the atomic number, which ranges from 0 (neutron) to 92 (uranium) and possibly, though not probably, even higher in unstable elements. But for most atomic numbers there are several atomic weights. The elements are not composed of like atoms, and atoms of the same mass do not necessarily have the same nuclear charge, and therefore the same properties. Aston has shown that a chemical element, like a Linnaean species, is something far less definite than was once believed.

But above all Rutherford showed that even the elements have a history. Some are extremely stable, others have mean lives of a fraction of a second. But none are unchangeable. The familiar atoms are merely the survivors of a probably far greater number of unstable types. Since the discovery of the neutron in 1932 over a hundred unstable types of atom have been produced, and the number increases every month. Rutherford has done for elements what Darwin did for species.

The process of radio-active change is still little understood, but it is abundantly clear that the nucleus of a radio-active atom breaks up, not because of any external agency, but through its internal instability. This appears to be so, even when the change is produced by bombardment with neutrons or other fast particles. The projectile enters the nucleus, which then breaks up after an average period which varies from a fraction of a second to some weeks. But the breakdown is an internal breakdown.

Just as the elements turn out to be mixtures, so do many compounds which were thought to be pure. Liquids often consist of several different molecular types in equilibrium. And the vast majority of known compounds are unstable, or to use a more exact phrase, metastable. Most organic compounds spontaneously rearrange themselves. Some, such as methyl (CH_3) whose existence has only recently been proved, are so intensely reactive that at ordinary temperatures they only last for a very small fraction of a second. Others have a survival period of many years under suitable conditions. But the importance of conditions is shown by the fact that many compounds which are fairly stable in glass bottles will not last in quartz bottles, which admit ultra-violet radiation.

The laws of organic chemistry are not absolute laws, they are functions of the human time scale. If men lived a million times quicker they could work with methyl and other compounds which we describe as radicals. If we lived a million times slower most of organic chemistry would be beyond our grasp, so unstable are the substances concerned. Time is at the very heart of chemistry, and just because industrial chemistry is concerned with change, the study of reaction velocities is assuming a greater importance each year.

And here the central problem is probably that of unimolecular reactions. Why does this unstable molecule break down in an average time of half a second, and this one in ten years? Why is this molecule stable in watery solution, but unstable when united with an enzyme? The answers can be given, where at all, in terms of quantum mechanics, with its strange union of chance and necessity. Nowhere can we say, "this molecule will inevitably rearrange itself within a certain time." We can only state the probability of such an event. To quote Engels once more:

> "One knows that what is maintained to be necessary is composed of sheer accidents."*

* F., p. 55.

And the calculations which give us our chance and our law are based on quantization, that is to say on the fact that energy is transferred in definite amounts. This principle is probably to be regarded as the simplest and most general form of the law of the transformation of quantity into quality.

Before I leave the subject of chemistry, I should like to point out the very beautiful principle of resonance energy, mainly due in its detailed developments to Linus Pauling. When the electronic theory of valency was put forward, there were many doubtful cases. For example, should carbon monoxide be written $:C::\dot{\underset{\cdot}{O}}:$ with two pairs of electrons shared between carbon and oxygen, or should it be written $:C\vdots\,\vdots O:$ with three pairs shared between them? Those two hypotheses would give somewhat different values for the distance apart of the atoms in the molecule, for the heat of reaction, and so on. And there was a certain amount of evidence for each of them, but neither fitted very well. Pauling's answer to the question was, "both." The molecule oscillates between the two forms, and that oscillation gives rise to an extra term in its energy which may in some analogous cases be very considerable.

I need not give further examples of the phenomenon of tautomerism, where several formulae for the same molecule are possible. Tautomerism is one of the properties of the aromatic and semi-aromatic compounds, and has long been familiar to organic chemists. But whereas tautomerism was originally regarded as a blot on the fair face of organic chemistry, it now turns out to be a property which, when established, gives us more, not less, information than we had before. The whole matter is a beautiful example of dialectical thinking, of the refusal to admit that two alternatives which are put before you are necessarily quite exclusive. Unfortunately, time is short, and I must pass on to biology.

4. *Biology*

Engels was a very thoroughgoing materialist in some ways. He did not believe in any impassable gap between chemistry and life, in spite of the fact that Pasteur had shown that all the supposed cases of spontaneous generation of life in putrefying materials were really to be explained by infections by air-borne bacteria, moulds, and the like. Even these bacteria are fairly highly-organized living beings, certainly with millions of years of history behind them, and the most primitive living beings, he thought, would probably be somewhat smaller and simpler. He went farther, and said that life in its simplest form was the mode of existence of proteins (or albuminous bodies, as they were then called). That seemed rather a rash statement at the time, because all the living things known when he wrote contained hundreds of different compounds besides proteins. In fact, until a couple of years ago, it seemed as if Engels had made rather a bad mistake, and let us admit that he did make several mistakes—he backed the wrong horse on at least three occasions in scientific disputes. In the last two or three years the gap between chemistry and life has been bridged to a considerable extent by the study of viruses. The agents of certain diseases will pass through filters which stop ordinary bacteria. Some of them are of a reasonable size, though too small to see with a microscope, and are stopped by membranes which will let through coloured particles like the molecules of haemoglobin, or even of haemocyanin, which are much larger. Others can pass through any membrane which will let through a large protein molecule. Some of these latter viruses have proved actually to be fairly large protein molecules. The most striking is the tobacco mosaic virus, which was first isolated in a rather impure state by Stanley in America, and of which several different species have since been studied by Pirie, Bawden, and Bernal in this country. These viruses may also affect potatoes and other related plants. From a diseased tobacco plant one can extract considerable quantities of the virus, which is a protein. One can purify it and keep it indefinitely. It does nothing whatever, provided it is kept completely aseptic. It is just a protein of the group called nucleoproteins, with rather peculiar properties when dissolved in water, but nothing to suggest it is alive in any way. One injects a little into a tobacco plant. A disease starts, the plant gets covered with yellow spots and wilts away. After two or three weeks, one can extract many thousand times the weight of protein which was injected.

One can then ask the question, is this nucleoprotein

alive and reproducing itself, as we say that a louse which irritates the skin is alive? Or, alternatively, does it somehow persuade the plant to copy it as, for example, a limerick persuades a man to repeat it? It may be that this question has some meaning. It may very well be that it is an example of the kind of illegitimate question against which we are continually coming up in physics; the kind of question which cannot be answered except verbally, because nobody can suggest an experiment to decide between the two alternatives. No one can really say that it makes any practical difference if you plump for one view rather than the other. In other words, there is a good deal to be said for both of the answers. We naturally do not understand the details of the reproduction of this virus. (I use the word "reproduction" because it is employed both for the reproduction of a photograph and of a living organism, and is therefore neutral.) It will take generations to work it out. Nevertheless, the bare fact that some people argue with fury that such a virus is alive, and others argue with equal fury that it is not alive, while some say there is no sense in making a distinction, indicates that not by any theoretical discussion, but by actual chemical technique, we have got the beginnings of a bridge between life and chemistry, even if we do not understand in the very least how a virus actually multiplies.

What is a protein? If we digest it and break up its large molecules with acids or enzymes it yields amino-acids. An amino-acid is a very peculiar stuff. It is a beautiful example of the union of opposites—an acid and a base. It displays some of the properties of both and also novel properties. A typical amino-acid is alanine:

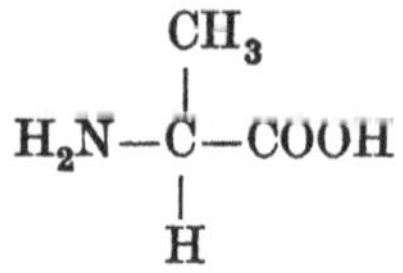

It has an acid group, a basic group, and a third radical, in this case methyl, attached to one carbon atom. It does not, in general, migrate in an electric field as a base like methylamine or an acid like acetic acid does. On the other hand, it does orient itself in an electrical field, because it ionizes both acid and basic groups, becoming:

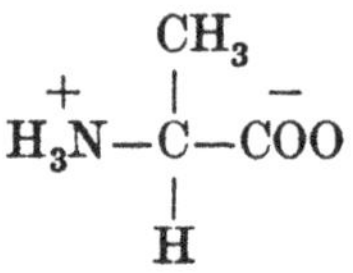

and therefore it lowers the dielectric constant of water considerably, and has a number of other curious properties. The union of a number of amino-acids according to a particular pattern gives you a protein, which again manifests novel properties, which are gradually being explained in terms of its atomic and energetic structure.

Every virus so far examined, and also bacteriophage (a group of ultramicroscopic agents which destroy bacteria) has been found to be a nucleoprotein, that is to say a protein combined with nucleic acid which is also an ampholyte, as a molecule with both acid and basic groups is called. I think that Engels would have been remarkably pleased to find that in the chemistry of proteins, the physical basis of life, the union of opposites is a condition for the emergence of novelty.

With regard to the general characteristics of life, it

appears to me that in the present state of our ignorance, our only satisfactory way of looking at it is as a union of opposites. For example, the living organism, as has been realized for a long time, is not something just there, but something that happens. Each living thing has its own way of life. A French biologist writes that it is inadequate to say that a dog lives and a rose lives, we ought to say a dog dogs and a rose roses!

Among the most universal manifestations of life are anabolism and catabolism—that is to say, the building up of complicated organic compounds, and their breaking down. In simple organisms, these are not closely correlated. You may find growing micro-organisms which utilize as much as 60 per cent of the energy of their food for growing, or you may be dealing with a flat-worm which you can leave for many months until it has shrunk to one-hundredth of its former size, and then if you give it food, it starts growing again. In these simple forms the unity of opposites is conditional, as Lenin put it. But in the higher animals there is a very close balance between the two, particularly so in man. For example, in a growing child Rubner found that only about 5 per cent of the energy in its food is actually stored away for growth. This is an average. It is much higher in babies, and lower in active children. In other mammals the figure is about six times higher, but there is always a fairly close balance between the two, and in the adult, of course, the balance is pretty well complete. This particular unity of opposites is part of the very nature of the higher animals, though not of the higher plants, nor of many lower animals.

We now pass to a great question which still agitates biologists, although the philosophers claim to have decided it long ago—"is a living organism a machine?" The mechanists say "yes." The vitalists say, roughly, that it is a machine plus a soul, or vital force, which constantly interferes with the mechanical working. The organicists speak of a unity of a special kind—a unity of the organism such as is not found in machines.

We can agree with them up to a point; but still, I think, we can say rather more than that. We have to ask ourselves, what do we mean by a machine? I think we mean a whole whose behaviour can be fully explained by the individual properties of its parts; and further in a great many cases we mean that a part can be removed and another actually substituted for it. That is not always the case, but it is a very common property of machines that spare parts are of value. From one point of view, the opposite to a machine is an individual, which, of course, etymologically means something which cannot be divided. A well-known example of this definition was Humpty Dumpty. You will remember that when Humpty Dumpty had his great fall, "All the king's horses and all the king's men couldn't put Humpty Dumpty together again." Whereas with an ordinary machine, unless it is broken too seriously, it is possible to put it together again, perhaps with the help of a few spares.

Now it is probable that from the point of view of modern physical theory both a machine and an individual are abstractions. There is no such thing as a 100 per cent machine or a 100 per cent individual, because we cannot actually isolate any system completely from the rest of the universe. Therefore if our knowledge of a so-called machine is sufficiently accurate, there will always be something unexplained by the internal properties of the system we are studying. That is one form of the uncertainty principle. Again, you do not seem to find 100 per cent individuals, even in the case of electrons. The form of the uncertainty principle to which I referred on page 82 implies that this is bound up with the fact that you cannot label them or distinguish them in any way except by constant interference.

Now it is part of the very nature of an organism to interact with its environment. It not only adapts itself to the environment, but to some degree it adapts the environment to itself. The latter activity is most strikingly displayed in the activity of man. Further, there is no sharp line between an organism and its environment. For example, few physiologists would claim that food in the stomach was part of the living system, but they would be considerably more doubtful as to whether the sugar from the digested food was really part of the living system when it had passed into the blood-stream. And after it had been stored away as glycogen in a cell most people would say it was part of the living matter. There is no particular point at which you can draw the line.

Let us see what happens if we say "A living organism is at once a machine and an individual." As soon as we do that, we stop asking whether man is an individual, which is a metaphysical sort of question. We ask, how much of an individual is he? Is he more so than an apple-tree or a frog? As soon as we ask that question, we begin to get a very interesting set of answers.

If a man were completely an individual, he would have no spare parts, so to speak. But he has some replaceable parts. If you lose a pint of your blood, you can replace it by a pint of mine. If your pancreas ceases to function properly and to manufacture insulin, so that you get diabetes, you can replace it to a considerable extent by insulin derived from the pancreas of a pig; though you have to exercise considerable caution about the amount which you put into your blood, whereas your pancreas manufactures the right amount without your thinking about it. It is quite certainly established, on the other hand, that a leg does not usually function as a spare part. You cannot in general graft one on from one man to another, or even from one mammal to another, unless they are genetically very similar. Further, if you cut a man in two, either one part dies or both parts die. If you cut him at a finger joint, the finger tip dies; if you cut him at the neck, both parts die. Whereas you can cut many plants and some worms in two in such a way that both parts will live. You can graft together frogs, not merely of different parentage, but of different species, provided you do it at a sufficiently early stage in their life history What is more, you find that although an adult man certainly cannot divide in two, a sufficiently early human embryo can do so, producing a pair of monozygotic twins. As soon as we look at the thing from this point of view, we conclude that on the whole there has been a progress towards more complete individuality, both in the development of the individual and in the evolution of the race. This is the kind of help you get to your thinking if you once admit that these apparently opposite statements about our individuality have a certain amount of sense in them.

Modern physiology begins with Descartes' theory that animals are machines, while men are machines with the addition of a soul, which he located in the pineal gland in the middle of the brain. There was plenty of materialism in the ancient world, but it did not take the form of asserting that animals were machines. For almost all the machines in the Roman Empire, such as carts, mills, and galleys, were worked directly by human or animal power, though occasionally, as in the bow and catapult, the power was stored up for a time. But by the end of the Middle Ages far more complicated machines were in existence, such as windmills, and above all clocks, which work continually for long periods with stored energy. The theory that animals were machines therefore became possible, and as soon as it became possible to propound it without risking one's life, Descartes did so. In the case of man only, the soul was superadded.

The eighteenth-century materialists discarded the human soul; and if modern brain surgery has not completely disproved its existence, it has at least shown that it is possible to remove the pineal gland without destroying consciousness and will. Now the machine theory of man works well up to a point. You can treat the heart as a pump and study the mechanism of the circulation. You can treat the eye as a sort of camera; and actually, when the optician prescribes new spectacles for you, he is treating the eye as an imperfect piece of machinery which he can improve. Again, our knowledge of the main chemical events which lead up to muscular contraction is probably by now fairly nearly, though not quite, complete. Above all, the laws of energy transformation which apply to non-living systems also apply to animal bodies and to the human body as a whole, with a very high degree of accuracy. This is the reason for the fact, which Lavoisier regarded as so unfortunate, that people who have to work hard need more food than those who do not have to work, although, as Lavoisier pointed out, they generally have less money to buy it with than people who do not work so hard.

I believe that our analysis of a living organism is only likely to be satisfactory if it starts off with a recognition that this mechanistical interpretation is of very great value, that Descartes did a very great thing for physiology, and that although we need not go all the way with the eighteenth-century French materialists, we still have something to learn from them.

I can perhaps illustrate this from my father, Professor J. S. Haldane's approach to the problem. One of his early books, *Life and Mechanism,* was devoted to a criticism of the mechanistic point of view, but in the first chapter he gave an account of that point of view which was so satisfactory that it was translated and reprinted almost verbatim in a German periodical which supported the mechanistic theory. Unfortunately, hardly any mention was made of the subsequent pages, in which he pointed out its defects! For when one goes a little farther, one finds that even if one is a rigid mechanist in theory, in practice one needs another principle. One finds that the organism somehow regulates itself. It tends to go back towards normal after any disturbance. Each detail in that regulation can perfectly well be interpreted mechanically or chemically, provided that the background is somehow kept normal. For example, the arterial pressure is kept fairly steady in a man who is not taking violent exercise. It is kept steady by the working of pressure gauges which are situated in small organs attached to the aorta in the chest, and to the carotid arteries in the neck. From these end-organs nerve fibres pass up to the brain. If the pressure rises unduly, large numbers of impulses pass up these nerve fibres. They can be detected quite well with suitable apparatus. When they get to the brain, there is a reflex action. Impulses pass down the vagus nerve to the heart. The pulse rate slows down until the blood pressure comes back very nearly to normal level.

So far, there is nothing different in principle from what you find in any self-regulating machine, but all these elaborate mechanisms only work provided, among other things, the chemistry of the blood is normal. If the amount of carbon dioxide or oxygen becomes sufficiently abnormal, the mechanism for blood-pressure regulation will certainly break down; in order for them to be kept normal the activity of the lungs must be regulated by another mechanism similar to the one for regulating the blood pressure. In order for the lungs to function normally, the kidneys have got to keep the chemical composition of the blood normal in other respects. The adrenal glands must not produce too much adrenalin, or the blood pressure regulation will be interfered with; the pituitary gland must secrete pitressin in the right amount if the kidneys are to work normally; and so on. In fact, one may say that this part of the mechanism works all right provided its background is normal. It is important to realise that that normality, in the long run, implies not only internal events in the organism but a relationship of the organism to its environment. A man must have a proper amount of oxygen in the air breathed, which he will not get if he goes into sufficiently ill-ventilated mines. He must have an adequate food supply, for which he must either work or induce other people to work for him.

There are some people who say that this merely proves that man is a complex machine. I am the last to deny that he is one. It seems to me, however, that in practice the physiologist, although he may be and should be mechanistic in his details, is never mechanistic about the organism as a whole. And the basic principles of physics are not of such a nature as to force him to the view that because an organism in its details observes physical and chemical laws, therefore it must be a machine in the sense defined above. On the contrary it is possible, without denying the validity of physical laws, to adapt biological theory to biological practice by saying that the organism is something more than a machine, in the sense that it has a unity of a type which the machine lacks.

Many physiologists found my father's theories rather difficult to follow; but they found that his practice was fairly easy to imitate, given sufficient technical skill. He was continually reproached with having given a purely chemical account of the behaviour of the blood as a carrier of gases, based on the behaviour of blood exposed in bottles to various gas mixtures, and having switched over from that to deal with the regulation of breathing as a vital phenomenon not to be explained on the principles of physics and chemistry as then, or even now, understood. Nevertheless that method of attacking the problem did, as a matter of fact, work pretty well. He insisted particularly on the uselessness of vitalism, the theory that the organism is a machine plus vital force; and he believed that in the long run physiological principles would be found to hold in non-living matter also. So he welcomed the dynamical theory of the atom as a self-repairing system as being the beginning of such an explanation.

Now if anything like that is true, if it is true that we may expect to explain life, not as chemistry plus X, but as chemistry with certain principles which are not important in ordinary chemical practice emphasized to a considerable degree, it is possible that chemistry, like physics in the past, will develop serious internal contradictions, and that it is not the biologist, but the chemist, who will do most towards bridging the gap. It is perfectly clear that I cannot suggest in any detail the lines on which such a contradiction will develop. It may be that systems much larger than a molecule, and in particular, whole cells, may prove to have a unity such as is found in a molecule, and a system of energy levels peculiar to the cell as a whole, just as the molecule has systems of energy levels which do not belong to any of its constituent atoms, but yet are not imposed upon it from outside.

Now from the point of view of physiology, as developed by most physiologists, the organism is something which repairs itself and brings itself back to normal if it is not too far injured. If it is altered beyond a certain point it may patch up some sort of compromise, such as a scar or a "by-pass" to a blocked blood-vessel, or it may die. In general, it attempts to restore its original structure or function; and furthermore it reproduces its like. It is well known that the effects of ordinary environmental changes are not inherited. If that were so, a very large number of children would have been born in the last twenty years with missing legs and other reminders of war wounds. It is worth noting that a perfect organism with those properties would be incapable of evolution. It is only the failure of organisms to come up to this standard which has made evolution possible. It is obvious that the very first requisite of evolution is death. Unless some organisms died, there would be no room for others of a more perfect character to replace them. Furthermore, in certain species there is no natural death. For example, plants like some tropical trees which can go on growing for centuries, presumably evolve a good deal more slowly than species which have a more frequent alternation of life and death.

Here I would add a brief linguistic note. We all vaguely realize that life and death somehow form part of a whole. We may attempt to state this in poetry or in religion. Yet we have no word with which to express the unity which they form except the word life, just as we have no word for day-plus-night. We use the word day both to mean the illuminated part of the twenty-four hours, and the whole twenty-four hours. Nevertheless, in a better developed language we should, I think, realize that just as a day and night form a unity, so do life and death.

I am not going to go into the historical evidence of evolution, but I shall briefly discuss how it occurs. Darwin, you will remember, believed that it took place not only as a result of natural selection, but from the inheritance of the effects of use and disuse. Darwin did not hold the latter view so strongly as Lamarck and Herbert Spencer, but Engels followed him in his partial Lamarckism. That view is almost completely unsupported by serious evidence. It seems as if each organism represented, except for a special case which we shall consider later, a fresh start.

Darwin took variation for granted, and thought that all variations were inheritable. This was largely due to Cobden and the repeal of the Corn Laws. Darwin took his evidence mainly from animal breeders. On the other hand his contemporary in France, de Vilmorin, showed that in a plant such as wheat, where self-fertilization over many generations is possible, you can arrive by prolonged inbreeding at what is now called a pure line, a group of plants in which all the individuals have the same properties in so far as they are determined by heredity, and hand down the same hereditary qualities to their offspring, and within which such differences as exist are due to environment and are not inherited.

In England, of course, wheat breeding on a large scale came to an end with the repeal of the Corn Laws, and was not resumed until the twentieth century. No pure lines of animals were produced in any country until about 1915, and no pure line of agricultural animals has yet been produced. Pure lines are mostly confined to such small animals as rats and mice.

It is clear that if Darwin had had the knowledge which we now have of pure lines, he would not have taken the view that variation due to the environment was generally inherited. He would have been forced not to take variation, or rather heritable variation, for granted, but to search for its origin.

When I said that a pure line bred true, I was not speaking the strict truth. The vast majority of individuals show the characteristics of the line with considerable accuracy; but in a sufficiently large population, you will find some which diverge from it. You will find that these novel characteristics are generally handed down to the offspring. The change by which a new heritable variation arises is called mutation. It is, if you like, the opposite of heredity; it negates it. We now know that the basis of heredity is to be found in what are called genes, which are units located in the chromosomes of the nucleus of the cell, units which, like viruses, either reproduce themselves or persuade the rest of the cell to produce them, and which, like viruses, are composed of nucleoprotein; though as virus nucleoprotein does not give the Feulgen reaction, they are nucleoproteins of different kinds.

The history of the gene theory* furnishes a beautiful example of the dialectical development of science. Mendel, who discovered some of the basic principles of genetics, called a gene a "*differendierende Merkmal,*" that is to say a character which behaved as a unit, for example, yellowness as opposed to greenness of the cotyledons in peas. He did not apparently think of it as a material object. It was found that a gamete (e.g. a spermatozöon, pollen grain, or egg) carried one set of genes, all (except in special cases) different. The union of two gametes gives a zygote, such as a man, fly, or pea plant, with two sets. Corresponding genes in the two sets may be alike, or they may differ. If they differ, in which case the organism is called a heterozygote, it produces equal numbers of gametes with the two different allelomorphic genes. Thus a cream guinea-pig produces equal numbers of gametes with genes which, when they meet their like, give rise to whites and yellows respectively.

* An elementary account of some parts of this theory is given in Chapter 2 of my *Heredity and Politics*, 1938. For a fuller account see Morgan's *The Theory of the Gene*, Oxford, 1926.

Bateson called genes factors, a neutral or positivistic word which could have meant either material particles, or qualities such as modes of vibration of the nucleus as a whole. Later, he and Punnett took the more materialistic view that recessive factors were simply the absence of dominant factors, and in 1902 Correns, developing an earlier suggestion of Strassburger, suggested that they were material particles strung out in a line along a chromosome.

About 1910 Sturtevant showed that this theory worked extremely well for the fly *Drosophila*. Chromosome maps were made for this and many other organisms. They were based on the principle that genes in the same chromosome which have come from the same parent tend to go out into the offspring together. If they have come from different parents they tend to go out separately. Thus if a pair of rabbits are derived from a black father with white fat and a white mother with yellow fat, most of their children who are white will also have yellow fat. But if the black father had yellow fat and the white mother had white fat, very few of the children will combine white fur and yellow fat. There is no such tendency in the case of genes in different chromosomes. Thus in the case of rabbits it makes no difference to the progeny whether the genes for yellow fat and long hair have come from the same or different parents.

The greater the tendency of two genes to stick together in this way, the nearer they are on the chromosome map. Thus if they only separate once in a hundred times they are said to be one unit apart. If they separate five times in a hundred they are five units apart. Chromosome maps made on this principle always proved to be lines, not for example, circles or figures of eight. Their validity was proved when Stern showed that a microscopically visible break in a chromosome involved the separation of two sets of genes as predicted by the map, and Painter was able to locate some of the genes in visible structures in certain very large chromosomes. In fact the genes could be treated as points on a line. Clearly this cannot be strictly true, for the gene is a structure composed of atoms and a point has no magnitude. However, the theory broke down, not under external criticisms of this sort, but under internal contradictions.

Some genes (e.g. bar eye in *Drosophila*) turned out to be duplications, that is to say whole sections of chromosome represented twice over. Others, such as notch (on the wings) were found to be deficiencies, that is to say losses of a small section of a chromosome. Still a third set were shown by Muller and Prokofieva to be due to inversions of a small section of a chromosome, i.e. the substitution of an order *abedcfg* for the original *abcdefg* of the genes. But still it was maintained that "real" genes differing from the normal type were due to "point mutations."

Goldschmidt originally held the over-simple theory that various genes allelomorphic to one another (e.g. those for orange, yellow, and white fur in the guinea-pig) only differed as regards quantity of gene substance. This was disproved by Wright, though it had been a very fruitful theory. Goldschmidt has recently switched over to an extreme theory of another type, namely that all genes differing from the normal are merely rearrangements of the standard type of chromosome, and that it is therefore incorrect to speak of genes in the wild or standard type of a species.

This is a very interesting negation. Sturtevant showed that the genes were arranged in a line. Muller and Prokofieva found that a reversal of part of this line constituted a new gene in certain cases (though rather rarely). Goldschmidt argues that therefore genes are an illusion. This type of negation, which may be compared to sawing off the branch on which you are sitting, or casting out devils by Beelzebub the prince of devils, is rarely, if ever, wholly successful, but is generally fruitful. Berkeley used it when he considered the structure of the eye as a material system, and deduced that we have no direct knowledge of matter, and no reason to believe that it exists. The actual result has been a correction of our ideas concerning matter rather than a disproof of its existence.

The probable result of Goldschmidt's critique will be the conclusion that a gene is an organ, a part of the chromosome with a definite function, extending over a finite region of the chromosome, and usually behaving as a unit in the formation of gametes. But the gene will not be regarded as a point, and it is even possible that different genes will be found to overlap. "Knowledge is the eternal infinite approach of thought to the object. The expression of nature in man's thought must be understood not in a dead abstract way, not without movement, not without contradiction, but in an eternal process of movement, of the springing up of contradictions and their solution."*

A whole chromosome or a gene can be altered by X-rays or other means, and, at least if the alteration is

* Lenin, quoted by Kemp in *Diderot, Interpreter of Nature,* by J. Kemp. (Lawrence and Wishart, London, 1937.)

within certain limits, the altered structure is reproduced. It is also probable that the process of reproduction of a gene occasionally fails. The change in a gene, whether direct or through faulty copying, is called mutation, and takes place spontaneously though very infrequently. Rare as it is, modern geneticists believe that mutation provides the principal raw materials for evolution.

Each woman in my audience possesses, I hope, in almost every nucleus of the cells of her body, two genes which are concerned with normal blood clotting, and each man possesses one such. Approximately once in every hundred thousand life cycles the process of reproduction of these genes breaks down, or the genes alter at some other period in the nuclear cycle. If a boy baby receives such an altered gene, he suffers from haemophilia, a condition in which the blood does not clot. There are probably from five to ten haemophilics in the city of Birmingham. Haemophilics generally die young, and thus natural selection keeps the frequency of haemophilia down. There is an equilibrium between mutation and selection. Now it is fairly obvious that haemophilia is disadvantageous, that it shortens life and diminishes effective fertility. I should like to point out, however, that even such an abnormality, though disadvantageous in general, might be useful under different circumstances.

Let us speculate for a moment and suppose that, ten thousand years hence, most people who have reached the age of five hundred need artificial hearts, as most old people nowadays need spectacles. In that case, rapid blood clotting might be a considerable handicap, Haemophilics might be at a considerable advantage over persons with normal blood. Of course, this is a somewhat fantastic and improbable example; but it is important to realize that cases of this kind are known. For example, Darwin pointed out that a large number of insects living on small oceanic islands have no wings, or very greatly reduced wings; although related species on the mainland have wings. Wingless forms have appeared by mutation in several insect species, and l'Héritier did the rather trivial but most illuminating experiment of keeping a mixed population of winged and wingless insects in a bottle constantly supplied with food. If they were shut up, the wingless ones soon disappeared, being less long-lived and fertile than the normal type. If they were allowed to escape, the winged forms soon vanished. They flew or were blown away and never returned. One can therefore produce a model for this particular type of natural selection.

Many species are full of recessive genes, that is to say, genes which are, so to speak, below the surface in ordinary individuals and do not show up unless two come together contributed by different parents. For example, populations of flies living in the middle of a continent may include one individual per thousand carrying the gene for winglessness. If so, one fly in four million would be wingless, and probably less likely than the average to leave progeny. Such genes constitute a reserve which might save the species were it transferred to a windy island. But in general, mutant genes diverging from the normal are undoubtedly harmful to the species in the existing environment, and many of them would be harmful in any possible environment. Albinism, for example, is disadvantageous in any environment where the eye is used. Even the Polar bear and the white fox have coloured eyes.

It is worth pointing out that mutations are generally harmful, not because they disturb the normal relations between an organism and its environment directly, as in the case of these genes for winglessness, or of genes for the absence of eyes, but because they disturb the harmonious internal relationships of the organism. For example, in plants, we find genes determining the overgrowth of the edge of a leaf, thus producing the type of crinkled leaf found in some cabbages, which is less effective for photo-synthesis than a flat leaf. In animals a rather comic example of misfit due to an inharmonious combination of genes is furnished by the cross between a bulldog and a basset-hound, which has a skin that would fit a dog twice its size. You see the same misfit to some extent in the head of an ordinary bulldog, where the skin is wrinkled. The genes determining the growth of skin and bone are largely independent; and if one varies when the other does not, disharmony arises.

Most evolution, as Herbert Spencer pointed out, is rather an adjustment of internal relationships than of external relationships. There can be little doubt that Darwin laid too much stress on the relationship of the organism and its environment, and too little stress on its internal relationships. It is worth noting that a change of environment in an animal or plant leads to internal contradictions, to failure of what had previously been harmonious relationships of the organ or process within the living organism. For example, a fish living in water has little need for a system of muscles and nerves to regulate its blood pressure. The blood which rises from the lower parts of the fish to the upper does not have to do any work in overcoming gravity. If the fish stands on its head, the blood does not tend to flow from its tail into its head. The moment, however, that the fish leaves the water and comes on the land, the moment it starts to change into an amphibian, it becomes necessary to have a particularly complicated set of muscles and nerves to regulate the bloodflow, and to prevent all the animal's blood running into its legs. It has to be much further developed in an animal such as man. Everyone knows what it feels like when it fails, when we first get up after lying in bed for several weeks. All the blood leaves our head, and we feel faint.

Man himself is still subject to a number of maladjustments of his anatomy and physiology, due to his assumption of the upright position. Our insides are suspended in a manner which is more suitable for animals going on all fours. As a result, we are liable to hernia, prolapse, and a number of other complaints. We do not normally regard hernia as caused by the fact that its victims do not walk on all fours; we put it down to internal maladjustment. The fact that a changed environment leads to these internal contradictions is very characteristic in biology.

Again, the adaptation of one organ may lead to less satisfactory functioning of the organism as a whole. As a result of the large growth of our brains, our noses have been turned round. If you consider the reflex of sneezing in a dog, the sneeze has a straight path from the lungs to the nostrils. In the case of man, it has to make a hairpin bend, and as a result we usually open our mouths when we sneeze in order to save our ear-drums.

Now many biologists find it difficult to believe that mutation is an evolutionary agency because its effects are, on the whole, harmful. A calculation suggests that as a result of mutation, fitness is reduced by anything from 5 to 10 per cent. It is possible that if we were able to do so, we should abolish mutation in man. If we did we should of course eradicate most congenital diseases; we should be healthier and possibly happier, but there would be no further possibility, if the theory which I am putting forward is correct, of the biological evolution of man. It is possible that some of the highly polyploid plants with large numbers of sets of chromosomes, like the great water dock, for example, have actually contrived to avoid the effects of mutation almost completely; and if this is so, they would seem to represent dead ends in evolution.

This point of view is entirely familiar to Marxists in the social field. Here is what Engels* says on the cognate social problem:

"He [Feuerbach] appears just as superficial, in comparison with Hegel, in his treatment of the antithesis of good and evil. 'One believes one is saying something great,' Hegel remarks, 'if one says that "man is naturally good." But one forgets that one says something far greater when one says "man is naturally evil."' According to Hegel, evil is the form in which the motive force of historical development presents itself. This, indeed, contains the twofold significance that while, on the one hand, each new advance necessarily appears a sacrilege against things hallowed, as a rebellion against conditions which, however old and moribund, have still been sanctified by custom; on the other hand, it is precisely the wicked passions of man—greed and lust for power—which, since the emergence of class antagonisms,

* F., p. 47.

serve as levers of historical development—a fact of which the history of feudalism and of the bourgeoisie, for example, constitute a single continual proof. But it does not occur to Feuerbach to investigate the historical role of moral evil."

On the other hand, biologists rightly see natural selection as a merely negative process. It cannot create novelty directly, though it certainly gives the conditions for its appearance. Hence there has been a considerable measure of reaction against Darwin, and a tendency to explain evolution by Lamarckism, by mere influence of the environment, by creative interference, and so on. I think that in the Soviet Union there is today a certain tendency to take the rather mechanical view of Lamarck that changes produced by the organism in response to a change of environment are directly inherited. This is by no means an illogical or *a priori* impossible view, but it appears to be untrue in the light of actual biological research.

It is worth pointing out that Engels happened to believe in it in about the same measure that Darwin did. In his unpublished manuscript notes for *Anti-Dühring*, he adopted the Lamarckian point of view to a considerable extent; and it led him to rather curious social consequences. For example, he thought that uncivilized men would probably be incapable of mathematical achievement and other forms of intellectual activity, because their ancestors had not practised them. He did not publish this passage. There were many points in it, and no doubt this was one of them, which did not satisfy him, and we may add at once that in the state of biological knowledge at his time, there was no way in which he could have reached what I believe to be the correct view.

In the violent discussions which have taken place on this question in the Soviet Union, some geneticists have been called anti-Darwinians because they have gone beyond Darwin in rejecting the inheritance of the effects of use and disuse. All the same, they have not so far as my information goes lost their jobs as a consequence. While, therefore, I am the last to suggest that all biologists in the Soviet Union are thinking as dialectically as they might, it is worth pointing out that the first work on what I believe can best be regarded as the dialectical opposition of mutation and selection has taken place in the Soviet Union, where it is still being carried on. The only other work in Europe along similar lines, so far as I know, is being done in my own laboratory.

If mutation and selection merely led to an equilibrium in the manner so far described, they would at best only serve to make a species more plastic and adaptable to changes of environment. The process would be of little interest to a Marxist. And so far it is only the equilibrium which has been directly observed under natural conditions. What follows is a deduction from the facts. It would seem that at least three types of true evolutionary progress are possible.

The first of these is that considered by Darwin. Some gene changes produce an exceedingly slight effect on the organism. Where this effect increases its efficiency in some respects without disturbing its harmony, the gene will be favoured by natural selection. But such selection will be exceedingly slow. However, it may account for some of the very gradual and apparently continuous changes which are revealed in the geological record. We shall see later that changes of a more rapid kind than Darwin postulated must also have occurred.

Secondly, each of two or three gene changes taken by itself may be harmful. But together they may be advantageous. Such cases have actually been observed, though we do not yet know the nature of the interaction involved. But we can readily see how a complicated organ like the eye could only be advantageously altered if a number of changes in it took place simultaneously. When, for example, the eye is lengthened without a diminution in the curvature of the cornea or the lens, short sight results.

The genes concerned in such a change will remain rare as the result of negative selection. But they will occasionally come together by chance. In a large population with little inbreeding they will soon separate again, and as each is only favoured when united with the others, the new type will not increase in numbers. There are, however, three ways out of this impasse.

If the population adopts the habit of self-fertilization or close inbreeding, the new combination of several favourable genes will not break up. Inbreeding, however, as Darwin showed experimentally, has other disadvantages. If small groups of individuals are isolated, each will have a chance to try out a different gene combination. The more successful groups will increase in numbers, and their members will migrate. They may exterminate the less fit groups, but generally they will merely interbreed with them, and thus gradually spread the new gene combination.

Or a rearrangement of the hereditary material may

take place so that the genes concerned all come to lie close together in the same chromosome. They will thus be inherited almost as a unit. Such chromosomal changes certainly occur in nature. But they may have further consequences. The crossing between two types with a different chromosomal arrangement leads to a certain amount of sterility, so that a change of this kind is a step towards the formation, not merely of a new race, but of a new species. And the chromosomal rearrangement may have an external effect (e.g. on the eye shape). Such effects have so far only been observed in flies, and never in maize, where rearrangement of the chromosomes is very common. But if they are common they are probably one, though only one, cause of the apparently useless differences between related species.

My colleague Professor R. A. Fisher has suggested yet a third effect of the antagonism between mutation and selection. Where a mutation is common, any gene which protects the organism against its results will be favoured by selection. Thus, he thinks, mutant genes which were originally more or less dominant, become first recessive and then inactive. Meanwhile the accumulation of protective genes causes the organism to evolve. Here at last we have a suggested cause for evolution which has nothing to do with the environment on the one hand, nor with any mysterious inner urge on the other. It is, in fact, a beautifully dialectical theory.

Nevertheless, I happen to be one of its two most determined critics, because it appears to me that, as stated by Fisher, it runs counter to certain facts. For example, mutant genes should be more recessive in outbred than inbred species, if this theory were true. But they are not. I mention this controversy in view of the widely held theory that acceptance of Marxism is an emotional cataclysm which completely ruins one's judgement. If only Fisher were a Marxist and I were not, this theory might perhaps be applicable in the case in question. As a Marxist, I hope that Fisher's general argument may have a wider validity than at present appears likely to me.

I think it is clear that the Marxist point of view leads one to look for these creative antagonisms in nature, and to investigate them when one finds them, but certainly not to accept them blindly. I have little doubt that the final result of our investigations will be to modify to some extent the views that I have put forward in this book, which only represent the conclusions which I have arrived at on the basis of the data so far available.

Marx and Engels criticized Darwin for taking contemporary economic conditions as a model for competition in the animal world, and still more did they criticize contemporary economic competition for not rising above the animal level. Actually, Darwin was undoubtedly influenced by his age in two distincts ways. In the first place, in view of the prevailing economic theories, and particularly the theory of Malthus, he was apt to regard the struggle for life as an external struggle, a struggle between individuals, rather than a struggle between each individual and its environment, and as a struggle for external adaptation, rather than an internal struggle. In the second place, naturally enough, living in a period where progress appeared to be steady, he did not realize the necessity for occasional leaps in evolution. He thought that it was a continuous process, but we now know that such cannot possibly be the case.

Let us consider the number of chromosomes in the nucleus of a wheat cell. They may be 14, 28, or 42 in different species. Within a species the number is constant in almost all individuals. (Not quite all, or evolution would be impossible.) It is quite clear that the evolutionary step which led from one group of species to another must have been a sudden one, particularly as we know that plants with an intermediate number between those mentioned above are generally abnormal, and if not structurally abnormal, are sterile. We can observe in nature the occasional sudden doubling of the chromosome number owing to an abnormal cell-division. And we can see how such a process goes on beside the slower changes due to the selection, as we now believe, of numbers of genes, each with a slight action on the organism, which explain ordinary Darwinian evolution. We know that in many genera, for example, in *Empetrum*, the tetraploid form is more resistant to cold than the form with only two sets of chromosomes (by "tetraploid" I mean an organism which has four sets of chromosomes in its nucleus instead of the normal two). Almost all the grasses of Spitzbergen are polyploids, i.e. have more than two sets of chromosomes. If we imagine a plant population living in a region where the climate is gradually getting colder, the following events would take place: while the change in the climate was slight, tetraploids would occasionally come into being, perhaps once in ten thousand or a hundred thousand plants. They would be under a slight disadvantage as compared with the normal, slightly less fertile, even though more cold-resisting. As the climate worsened, the tetraploids would have less and less disadvantage, and finally, at a certain stage, the two types would struggle against one another on a footing of equality. Then as the climate got steadily colder, the tetraploid race would replace the normal one except for a very occasional variant. That, we now believe, is a much more frequent type of occurrence than Darwin thought. A species passes from stability through instability to a new stable stage.

The main fallacy, it seems to me, in the social application of Darwinism, has been the attempt by its use to justify economic and racial inequality. I have dealt with that attempt in much greater detail in my Muirhead Lectures for 1937, which have since been published under the title of *Heredity and Politics*. The absurdity, as it seems to me, of that point of view, is revealed by certain facts. In this country, and in the majority of other countries for which we have statistics, the poor do as a matter of hard fact breed quicker than the rich. Therefore, from the Darwinian point of view, the poor happen to be fitter than the rich. The attempt to describe the heritable qualities of the rich as anything analogous to Darwinian fitness seems to be simply propaganda. Whatever these qualities may be, they are not of the character which, according to Darwin, distinguish the fit or favoured race from the unfit or unfavoured race. They might even fit them for preferential survival in a different state of society, as a pipless fruit tree is fitted for survival under human protection, but not in the wild state. Similarly, it is frequently found that the conquered breed more rapidly than the conqueror. This probably occurred in the various countries of Europe after the Barbarian invasions. It is difficult to prove in most cases; but we do know that in the last sixty years, since the

British conquered Egypt, the population in Egypt has increased very much more rapidly than that of England. Therefore, if you say that the Egyptians are an unfit race, suitable to be conquered, this is in flat contradiction of the Darwinian point of view of what constitutes fitness.

In view of such facts, it is very necessary to separate the biological approach to these problems from political and economic considerations if one is to think clearly on matters of politics and economics.

A good Marxist will expect that the Darwinian theory of natural selection should contain its own internal contradictions. Engels* described it as "the first, temporary, incomplete expression of a recently discovered fact." So far as we know at present natural selection seems to negate itself in at least two ways.

First, as I have pointed out,† so long as a species is sparse, the fitness of an individual will depend almost wholly on its success in coping with inorganic nature and other species. But if it exists in dense populations, and occasionally under other circumstances, fitness comes to depend largely on defeating other members of the same species. Thus in the case of animals where each individual or pair lives in a restricted territory, a gene favouring the habit of devouring young members of the same species will diminish fitness. For its possessors will eat their own young. But where they live in large aggregates this will not be so. The first few cannibal fish in a school of many thousands will rarely eat their own young. They will generally eat those of others, and will thus have more food to turn into eggs and spermatozoa. Thus

* Letter to Lavrov, November 12, 1874.

† J. B. S. Haldane, *The Causes of Evolution*, London, 1932.

a gene favouring cannibalism will tend to spread. And in so far as it becomes common it will depress the fitness of the species as a whole.

Exactly the same argument applies to polygamous species where the males fight for females. Such males commonly develop special weapons and instincts (or modes of behaviour, if a more neutral word is preferred) which fit them for mutual combat, but not necessarily for the struggle for food and against enemies of other species. They also tend to be much larger than the females, and it is probable that some of this increase in size is transferred to the females (i.e. that not all the size-genes selected are sex-limited). Thus the size, behaviour, and so on are shifted from the optimum configuration needed for the struggle with "external" nature, and the species becomes less fit.

In plants the same process is most strikingly seen in connection with pollination. Here a gene for excessive pollen production is likely to spread, even if enough pollen is already produced to fertilize every ovule. Probably the extreme specialization found in the pollination of orchids and many other plants is also to be explained on these lines. It seems highly probable that the steady increase in size, ending in extinction, which is found in many fossil lines, was a result of intra-specific competition. If so there has been competition between species for co-operative behaviour, those which did not show it being wiped out.

Another internal contradiction is pointed out by Elton.* When two species are struggling, and in par-

* *Evolution*, Oxford 1938. (See also J. S. Huxley's essay in the same volume.)

ticular when one eats the other their numbers fluctuate. Volterra, Lotka, Nicholson, and others have shown that equilibrium is impossible in many cases. The eaters increase till they reduce the numbers of the eaten and then themselves starve. After this the eaters begin to increase again. The Soviet biologist Gause* has studied these fluctuations experimentally on small populations. He finds that under his conditions it is usual for the eaters to exterminate their prey and therefore themselves. This will rarely happen over a large area at once. But it may happen occasionally. Elton thinks that as a result of this eaters, whether herbivorous, carnivorous, or parasitic, are less ruthlessly efficient than they would otherwise be. Moderation has a certain survival value. I believe that the theory of evolution is likely to develop on these dialectical lines, on a basis of Darwinism, but transcending the rather crude Darwinism which is still used to justify unrestricted competition, and which, like unrestricted competition, is really a thing of the past.

In passing from biology to psychology, we note that consciousness in many cases arises from a conflict of reflexes or habits. We are normally unconscious of breathing. A resistance or a heightened chemical stimulus makes us conscious of it, often unpleasantly so. We perform some simple task, such as knitting, unconsciously. It is only a change in the pattern or a break in the yarn which recalls us to consciousness of it. In order to do something out of the routine we must make an act of will, and consciousness is a preliminary to this. Samuel Butler developed this idea with great ingenuity and some perversity in *Life and Habit.* He even suggested that

* Gause, *The Struggle for Existence,* Baltimore 1934.

consciousness always arose in this way, and that this was its historical origin. I do not think that we know enough as yet to come to any useful conclusion on this question. But Butler's theory does not seem to be impossible, and although it gives no detailed account of the process concerned, it would, if confirmed, be in harmony with dialectical materialism.

5. *Psychology*

Man, and presumably other higher animals, has a mind; or, as others say, men manifest mind or are minds. The mind is the most amazing example that we have of the union of unity and diversity. Philosophers have occupied themselves, and will, no doubt, continue to do so, for many generations, with an attempt to explain how the various constituents of the mind are held together. From the Marxist point of view, the most obvious fact is that mind must be regarded as a process rather than a substance, though it clearly has a measure of unity and continuity. Now when we come to the historical theories of mind, we generally find that they involve a theory of a soul separable from the body. Further, the soul could be either wholly or partly separable. Some parts are, or might be, mortal, others immortal. The soul has been regarded by philosophers sometimes as a unity, sometimes as a plurality.

The theories are enormously complicated by the fact that the mind is commonly identified with that aspect or portion of the individual which is believed to survive death. This is, I think, more or less accidental. Actually there is plenty of evidence that the mind often dies before the body, and remarkably little that the body ever dies before the mind. We can roughly trace the growth of the immortal soul theory in Egypt. Although, of course, current religious ideas (except in the opinion of the extreme diffusionist school) go back to many origins besides Egypt. Immortality seems to have begun as a royal prerogative. The king's mummy was preserved. And the main economic effort of the state was directed to this end, at least under the fourth dynasty, the pyramid builders. But provision was made for the survival of other aspects of the personality, such as the ghost (ka), the shadow, and even the name. The practice of mummification spread among the well-to-do. But as the demand for immortality spread, other and cheaper means of assuring it were provided.

Under the Roman Empire a variety of mystery religions promised it to their adherents. And the means of attaining it were gradually cheapened, until they came within the resources of the poorest, even if the necessary rites involved going short of food and other things which might have prolonged life on earth. But if mummification was not to be a necessary condition for immortality, this meant that the "immortal part" must be the most abstract and intangible aspect of an individual, his mind. Christianity produced a somewhat different theory. The resurrection of the body was promised for the non-mummified. And the effect of ritual and other practices was not to avoid death, but eternal torment. However, this doctrine has proved hard to accept, and for one Christian who believes firmly in the resurrection of the body, probably five believe in the immortality of the soul.

The consequence of these religious theories has been disastrous for psychology. Philosophers have generally framed their theories of the mind in such a way as to be compatible with the religion of their time; and observed facts have played a subordinate part in framing them. We can imagine how the development of anatomy, physiology, and medicine would have been handicapped had the doctrine been widely held, and enforced by powerful vested interests, that survival of death depended on mummification. The situation with regard to theories of the mind has been analogous. In analysing them we must keep this central fact before us. Even those readers who are convinced of the immortality of the soul must admit that there are immensely powerful vested interests which stand or fall with this theory, and that the belief in it is very largely the result of emotional urges and of social pressure. When, however, we have made all allowances for the influence of religions on the theories of the soul, there is still a good deal more to be said.

At first sight it appears fantastic to suggest that not only the theories of the soul, but possibly even its nature, are to a considerable extent conditioned by the class struggle. I will try to justify this view. It will be remembered that, according to Plato, the soul consists of three different parts; a rational part; a spiritual part responsible for our more respectable but unintellectual behaviour, such as courage; and an appetitive part which is responsible for our less respectable instinctive behaviour, particularly for such appetites as greed and lust. Plato compared those three parts with a man, a lion, and a many-headed monster.

Now in the *Republic*, he explained the divisions of the soul on the analogy of the classes within the State. He compared them with the wise rulers, the soldiers, and the artisans and other mere workers at the lowest level. On the basis of which particular class in the State predominated, he described a number of different types of state, to each of which, he said, a human type corresponds. He described the kingly, or aristocratic, type of state, in which rule is in the hands of the very best and most intelligent people, and the corresponding type of man in which the intellect is master. Next came the "timocratic" man and state, ruled by the principle of honour, and exemplified by Sparta as a state, and an honourable but rather stupid soldier as a man. He went on to the oligarchic state, corresponding with the miserly man, and the democratic state corresponding with the flighty man of no very strict principles. Last of all comes the tyrannical state, in which rule is in the hands of one individual of base origin and little intellect and morals, to which corresponds the tyrannical man who is the slave of one of his lusts. It is very difficult to suppose that this theory of the soul was not actually influenced by the class struggle of his own time. At any rate he said that it was so influenced, and he was not a Marxist.

To mention a few other important thinkers on this subject, St. Paul followed the fundamental theories of Plato in many respects. Thousands of highly intelligent men have been occupied for the last nineteen hundred years in trying to determine exactly what St. Paul meant, and I do not pretend to have said the last word on this subject. However, his account of the soul is largely pluralistic, and was almost certainly influenced by Plato's, if perhaps indirectly. On the one hand, he writes of the lower part of the soul—the "lusts which war in our members," the mind of the flesh. Above this there was the soul proper with which he identified himself, and, in addition, by divine grace, there might be the indwelling Holy Spirit.

I should like to suggest that both Plato's and St. Paul's points of view were to a great extent influenced by the existence of slave labour, which led to a contempt for the body, the vile material body, and for the instincts connected with its functioning. This contempt might show itself in either of two ways. It might show itself in the view that the bodily passions were a nuisance, and that the best thing to do was to give way to them and get it over, and then go on with those intellectual processes which alone were worthy of a wise man. Alternatively, the Christians and some Pagan philosophers took the ascetic point of view. These base and fleshly tendencies had to be fought down as far as possible and suppressed. These attitudes, which made love, as we know it, a rare phenomenon, particularly in the ruling class, were a by-product of the contempt for the material world and for the less complicated forms of human activity which was inevitable in a state based on slavery. A reading of Plato's *Republic* will convince anyone that that social structure had at least a certain influence on these theories.

In the Middle Ages we find a unitary theory coming to the front. St. Thomas lived near the climax of mediaeval society, when it had reached a somewhat precarious equilibrium, and he put forward as satisfactory a unitary theory of the soul as has ever been given.

With the Reformation and the rise of the early capitalist class, of which the Reformation was a symptom, as not only Marx but such anti-Marxists as Mr. Belloc point out in great detail, psychological theory reverts to pluralism. I take Bunyan as an example, for the reason that although Bunyan was not a philosopher in the ordinary sense, yet for every one person who has read his contemporary academic philosopher, Cudworth, or for every ten who have read Locke, a thousand have read Bunyan. Therefore, Bunyan's ideas have done a great deal more than those of the philosophers to mould the ideas which the ordinary Englishman and woman adopt, or adopted till very recently, concerning the soul.

Bunyan never said that the soul consisted of a number of parts. Nevertheless, one of his most readable books, *The Holy War*, describes a civil war in the City of Mansoul between the supporters of Shaddai, King of Universe, on the one part, and the partisans of the giant Diabolus on the other. It is interesting to see how the descriptions he gives are a replica of the class struggle of his own time. To take a simple point—the people loyal to Shaddai are mostly plain "Mr.'s," for example, Mr. Conscience the recorder, a worthy old gentleman with a tendency to go to sleep. Whereas the peerage is extensively represented among the Diabolonians. You will find in one single catalogue, the Lord Fornication, the Lord Adultery, the Lord Murder, the Lord Anger, the Lord Lasciviousness, the Lord Deceit, and the Lord Evil-Eye, along with mere commoners such as Mr. Drunkenness, Mr. Witchcraft, and Mr. Heresie. There are one or two Lords on the side of God, but they are relatively rare. I want to suggest, therefore, that it seemed natural to Bunyan, in his own day, to use the contemporary class struggle as an image of what was going on in the individual soul; and even if he did not believe in a thoroughly pluralistic theory, like Plato, he found it natural to think along those lines because of what was going on around him.

The psychological theory and practice of "left-wing" Protestantism is based on the fact of conversion. The individual goes through a period of despair and repentance, followed by a sudden change of heart, which is ascribed to divine grace. But the state of grace is a somewhat uneasy equilibrium, in which various intellectual and emotional tendencies are stigmatized as doubt and lust, and ruthlessly suppressed. Conversion of this type was of course known in the Catholic Church, particularly among the saints. But it was only the extreme Protestants who made it the norm to which every individual was supposed to conform. I do not think that it is a coincidence that this religious tendency was a concomitant of the bourgeois revolution.

The ordinary Catholic sins, confesses, and repents periodically. There is seldom a violent or revolutionary change in his soul. And for this very reason he is often easier to get on with and has a more many-sided nature than the Protestant. The sanctified Protestant has suppressed one whole side of his nature. He is often a narrow but intense man, drawing energy from his internal contradictions, and very frequently devoting this energy with great success to the stern pursuit of money. He represents a splitting of the soul and a suppression of certain of its elements on lines determined by the contemporary class struggle. For example, sharp practice in business is infinitely less reprehensible to him than a sin such as adultery which strikes at the family and thus at the structure of society.

The main objections to Catholic psychological technique are not only that it is based on a false theory, but that it cannot adapt itself to changing industrial technique. The Middle Ages went down in a moral as well as a political collapse due to the rise of urban industry and the bourgeoisie. And today among the products of Catholic psychological practice are such characters as that of Franco and such events as the bombardments of Madrid, Barcelona, and Guernica.

In our own day we have the theories of Freud. One of his descriptions of the soul—I do not know if it is the latest one—presents it as consisting of the following parts, not, of course, entirely separate, but merging to a greater or lesser degree. First there is the Id, the irrational unconscious, which is connected with bodily desires of various kinds. Then follows the Ego, or fully conscious part of the soul, which is more or less rational, and concerned particularly with the relationships of the body with the external world. And finally there is the Super-ego, or Ego-ideal, more or less unconscious, and representing society, and above all, the parents. The Ego-ideal may be to a large extent irrational and tyrannical, giving rise to morbid feelings of guilt.

If we compare this philosophy with that of such a typical academic philosopher of the nineteenth century as T. H. Green, we notice striking differences. Freud's Super-ego corresponds to a large extent to what Green called the true self. Green identified himself with that part of the soul which was to a considerable extent in harmony with the existing society and with the example of his parents. Freud, who finds himself out of harmony with existing society, does not identify himself with that particular part of the soul. He regards it as a portion or aspect of his psyche distinct from the true Ego.

It seems to me that Freud's account of the soul is probably valuable from a therapeutic point of view. It is dialectical in so far as it shows the soul driven to acts, often of a strange character, by its internal contradictions, by the existence of repressed complexes with which it cannot deal by rational means, and by the struggle between life-instincts and death-instincts. Above all he has gone a long way to establish the great importance of the environment, even if he has probably ascribed too great an importance to the family as compared with society in general. But we may question the absolute validity of Freud's analysis. It may possibly be true for a large number of individuals in our existing society. It does not in the least follow that it was so in antique or medieval society, or that it will be true in a future and different form of society. In each case, there would probably be a measure of division within the soul. It does not follow, I think, that the division would always be on the same lines. I would suggest that the lines of division, and particularly those forms of mental activity which are regarded as reprehensible and pushed out of consciousness, are determined to a considerable extent by the existing society. I can quite imagine that in a communist society, children would be taught to repress various urges to accumulate wealth in different forms, which are quite respectable in our society; and possibly a psycho-analyst might have to bring such repressed desires to the surface before the psychology of a neurotic could be made normal. This may be an extreme example, but in view of the things which primitive societies find disgraceful, it is not necessarily absolutely fantastic.

On page 28 I quoted the passages*

> "The question arises, however, what is the content of the merely formal will—of the individual as well as the state—and how is this content derived.
>
> ". . . all the driving forces of the actions of any individual person must pass through his brain, and transform themselves into motives of his will in order to set him into action. . . ."

They show that Engels was quite aware of the fact of unconscious motivation, and would have welcomed its investigation. He would probably have preferred to describe the unconscious processes as cerebral rather than mental. But for a thoroughgoing dialectical materialist the exact point at which a line is drawn between the brain and the mind is, of course, a matter of convenience. At some future date it may be more satisfactory to regard events which Freud calls mental as physiological, just as digestion is nowadays regarded as lying more in the province of the chemist than of the physiologist. Or biology and psychology may become as inextricably fused as are some branches of physics and chemistry.

Marxism has extremely little positive to say about the

* F., p. 62.

individual mind. It is concerned mainly with the social relationships of the individual and the general materialistic point of view. A certain amount is, however, said about the general problem of the relationship of the mind to the physical world. The difficulties in imagining what that relationship is have been stated by Berkeley as clearly as by anyone else. Berkeley criticized matter as pictured by Newton and Locke, and his criticism is very largely valid. According to them, matter had extension, place, size, movement, and so on, but it had no secondary qualities, such as colour, sound, taste and temperature. That view was dominant in the physical theory of the eighteenth century, and even now is entrenched to a considerable extent in scientific thought.

Berkeley's criticism was admirably dialectical. Here is a quotation:*

> "Hence it is plain that the very notion of what is called Matter or corporal substance involves a contradiction in it."

Note that he did not observe that mind or spirit does so too. One would think that as a bishop with a cure of souls he might have observed certain contradictions in the minds of his flock. But it was left to Hume to treat the mind as Berkeley treated matter.

Again, Berkeley defined spirit as that which perceives. You will see what a one-sided definition that is. He refers elsewhere to the activity of spirit, but he does not emphasize it in his philosophical analysis. From the Marxist point of view you are unlikely to get anywhere if you emphasize the perceptive side of mental activity at the expense of the active side. A beautiful example of

* *Treatise on the Principles of Human Knowledge.*

Berkeley's dialectical method occurs in the dialogue of Hylas and Philonous, in which he describes the transition from quantity to quality as one feels an object which increases in temperature. As, for example, a piece of metal gets hotter, you experience a continuous transition from a sense of warmth to a sense of pain. A materialistic analysis by physiologists has cleared the matter up to a considerable extent. Three different sets of receptor organs in the skin, each with its own nervous connections, are concerned. One set is excited by moderate warmth. A greater heat also excites the end-organs concerned in the sensation of cold. And the feeling of heat is a compound sensation due to the simultaneous stimulation of both sets. You can often notice the feeling of coldness along with the feeling of warmth when you put your hand into sufficiently hot water. Finally you get the excitation of end organs causing a sense of pain, and that finally swamps all the others.

Now Berkeley regarded those contradictions, and such facts as the transition of quantity into quality, as a disproof of the existence of matter. They are, I think, a disproof of the existence of matter as Newton and Locke conceived it. He also says the following:*

> "Extension, figure, and motion are only ideas existing in the mind; one idea can only be like another idea. Therefore extension, etc., cannot exist in unperceiving substance."

That has undoubtedly been a very difficult criticism to answer. The Marxist point of view is as follows. The mind is a part of nature, and in particular it modifies the rest of nature, as well as being modified by it. Just

* *Treatise on the Principles of Human Knowledge*

because the mind is a part of nature, the processes which go on in it can be and are like the processes which go on in other parts of nature, and they do actually mirror them, although more or less incompletely.

"Ah," says the idealist, "you have admitted that your perceptions are like matter. If so matter is like your perceptions (or impressions, or whatever other word you use for your mental events). But you have direct knowledge of your perceptions and no direct knowledge of material objects. Your philosophy should therefore begin with perceptions, sensa, or whatever you choose to call them, and from them build up matter."

To which a Marxist might answer somewhat as follows: "My good friend (for I believe in your real existence, and do not think that I have built you up out of my sensations), I have three points to make against your argument. First, I believe that there was matter before there were sensations. This belief is based on an examination of nature, similar to that which has led you to the view that our knowledge of matter is indirect. For I think you reached that opinion after studying the human sense organs. And of two like things, one of which is model and the other copy, the one which has been in existence longest is the model.

"Secondly, even if our knowledge of external nature is indirect, we know enormously more about it than about our sensations. For our knowledge of it is social. I see my hand, and know that it contains a certain assortment of muscles, bones, and blood-vessels. This knowledge is based on the sensations of thousands of anatomists. I know the arrangement of the atoms in each hair (or at least roughly so). This knowledge is based on the sensations of Astbury when examining X-ray photographs. The socialized knowledge of thousands of people, even if indirect, gives more information than the individual knowledge of one. Robinson Crusoe would have had some excuse for being an idealist. I should have much less.

"Thirdly, I have to work in the real world. If I were a pure mathematician or a mystic, I might be an idealist. Though even they, unless they are completely selfish, communicate their ideas to their fellows through acting on matter by speaking or writing. If you really believe that you built up the world out of your sensations you have a very heavy responsibility, though I must say I congratulate you on some bits of it. However, I don't ask you to build a better world, but merely to help in changing this one. I think that the resistance with which you meet when you make this attempt will go a long way to convince you that you are up against reality. And do remember that your idealistic beliefs are very likely to lead you into the fatal theory that you can change the world simply by noble thinking and goodwill. If it were not for this danger I don't think I should trouble to try and argue you out of your position."

The question immediately arises as to the relationship between mind and brain. It is at present an unanswerable question, because we have not got the necessary data, and not in the least, I believe, because it involves contradictions or antinomies any more serious than those with which physicists are already dealing. For that reason Engels left it alone except for a few sentences. The additional knowledge of the last fifty years does not yet allow us even to pose the question in its proper form.

I am often tempted to ask, "What sort of thing is the

mind, considered as a physical object or process?" For since matter causes mental events such as sensations, and will, which is a mental event, affects matter, a complete physics (in its etymological sense of an account of nature) must include mind. However, it may be better to speak of mental events which are properties of material systems, provided we do not fall into the theory that mind is a mere epiphenomenon of matter.

Any suggestions towards a solution of this problem must be speculative. I am going to put forward my own speculations, based on scientific advances made during the last thirty years. They are in no sense part of Marxism, but they suggest the kind of hypotheses which a Marxist might reasonably investigate.

In the first place, we note that mental phenomena obey certain laws which are by no means analogous to the laws of classical physics, but much more like those which are revealed as governing the behaviour of small pieces of matter in the light of quantum mechanics. On the one hand, mental events cannot be accurately located in time—they fade in and fade out. That is a fact which has been known to psychologists for a long time. Similarly, they cannot be accurately located in space. I have no doubt that they are connected with the brain in some manner, because by influencing the brain in a very large variety of ways, I can influence the events in my mind. On the other hand, they do not seem to be located in any particular place; or at least some of them do not. It is true that a very small injury to a particular area of the occipital cortex at the back of the brain causes blindness to objects in some particular direction, but when we get to the more complicated mental activities, that is no longer the case. Lashley taught rats to find their way through a maze. He was able to destroy any small portion of their brains without affecting their ability. The ability, or memory, was not located in any special place. This is a very crude account of a very complicated series of experiments, and the criticism which could be levelled against this crude account could not be levelled against a completer account.

Another remarkable feature about mental events is the complementary relationship, as Bohr calls it. Just as we cannot determine at the same time the position and velocity of a particle accurately, so we cannot observe certain types of mental events at the same time. For example, we can be angry, but if we start thinking about it, as soon as we begin a detailed introspection we find that we are considerably less angry. This is a very trivial observation, but it means two things. In the first place, you cannot simultaneously observe certain activities of the mind; and above all introspection does not tell you accurately about the mental processes which are going on when you are not looking self-consciously into your own mind. You cannot be a mere observer of your own mind, any more than of material processes.

In the last ten years it has for the first time been possible to observe physical phenomena in the brain which are at any rate associated with mental processes which go on there. We find that various different sections of the human cerebral cortex, each of them composed of many millions of cells, are giving rhythmical electric discharges which involve the simultaneous electrical activity of many of these cells.

The most striking, generally described as the Berger

rhythm, occurs in the occipital cortex and is concerned with vision. It is an easy one to locate. We find that the rhythm is disturbed if we tell the person observed to look at something. It is altered if he shuts his eyes. It is changed if he undertakes any process involving visual thinking. Again, abnormal rhythms are found in epileptics and during sleep in normal people. These are new facts, and naturally they are under very careful investigation at the present moment, so that any attempts at a speculative theory based on them are inevitably three-quarters wrong, and may be 100 per cent wrong. Nevertheless, one is tempted to suggest the lines of thought to which they might possibly lead. The speculation would be that the mind, considered as a thing (or the physical fact which determines the existence of mind) consists of the resonance energy of these simultaneous discharges of living cells. If so, the mind would be an aspect of the unity of the body, as, for example, Aristotle and St. Thomas Aquinas believed; and just as molecules possess an energy structure which is not due to any one of their individual parts, so does the organism as a whole. Further, sensations, ideas and other mental events would be physical objects with a mass, although extremely small, and for that reason of very indefinite location.

And just as there is no absolute distinction between two electrons, there may be no absolute distinction between ideas or even sensations in different minds. This notion has long been a commonplace of idealist philosophers, who claim that rational discourse on abstract matters is only possible because two people can think about the same abstract idea, as they can perceive the same table or sun. I can see no reason why a materialist should not concur in this view. But in so doing he need not suppose that the idea has any existence apart from those who are actually thinking it at any given time.

Opponents of materialism point to the absence of purpose in inanimate nature, and regard this as a radical difference between mind and matter, or life and mechanism. The distinction holds good as between the higher animals and ordinary inanimate objects. If you have a number of mice in a house they will tend to congregate near a cheese. If you have a number of balls on a surface they will tend to congregate at its lowest point. But there is an important difference. A ball will not roll up a small hill in order to get the chance of rolling down a much longer one. But a mouse will make a detour, and go farther away from the cheese, in order to find a way round an obstacle. And a dog or a man will endure a certain amount of unpleasure or even pain for the sake of a future pleasure.

One of the most amazing features of quantum mechanical theory is the discovery that electrons and other elementary particles will leak through a potential barrier which they could never cross if the classical physics were true. The electron is imprisoned, for example, in a metal filament, and would gain kinetic energy, like a stone rolling down hill, if it could cross a gap to a positively charged plate. But to leave the metal it has to traverse a potential barrier at the surface of the filament and does not possess the requisite energy. According to the classical physics, it is like a stone in a small depression on a hill-side, which cannot get out so as to roll down the hill. There is no force acting on the electron or the stone which will take them over the barrier. But such an electron does get out, though the stone does not. Given the right data we can write down the probability that it will escape during the next second, though we can make no certain prediction for a particular electron. And on the basis of these calculations new types of radio valve are designed, since the statistical behaviour of millions of electrons is very accurately predicted.

The physicists do not say that the electrons can perceive what is going on at a distance from them and act on their knowledge. If they like a "model," that is to say a physical explanation, they say that the wave system of the electron leaks through the barrier, and the probability of finding it outside therefore increases with time. The fact is that, whether or not we take the wave system as a reality, the electron is influenced by surrounding objects in a manner not contemplated by physicists up till the last twelve years. We should be inclined to attribute perception and insight to it were it not that its behaviour, or rather the behaviour of a large group of electrons, is predictable mathematically, and animism has never been of practical value where accurate prediction was concerned. Indeed, it behaves less like a machine than do some animals which will persistently attempt to traverse an impassable barrier rather than make a detour. I suggest that the mind has physical properties analogous to those of a single electron or atom, properties which are lost in a group of trillions of atoms with the low degree of organization found in a stone, but which may be accentuated when the same number of atoms is organized in a mouse or a man.

Finally, we must realize that the mind, considered as a physical object, is a very strange one. We need not be surprised if it exhibits properties which seemed impossible to the physicists and chemists of former centuries. I do not see why a dialectical materialist should reject *a priori* the possibility of such alleged phenomena as telepathy and clairvoyance. I do not doubt that most of the reported cases rest on conscious or unconscious deception. But I can see no reason for regarding a certain lack of privacy in mental images on the one hand impossible, or on the other hand miraculous in the sense of involving a breach of natural laws of a very general validity. We should expect that such phenomena would be rare; for if they were common they would interfere with our normal perception and thought, and hence natural selection may be expected to have safeguarded us against them to a large extent. And when they occur we should expect them to occur, like quantum phenomena, rather spasmodically, and not with the certainty which characterizes the exactly predictable behaviour of large-scale material aggregates. I do not, of course, affirm that such phenomena occur. But if their occurrence should be proved, I do not think that this would disprove materialism, or even revolutionize science; though it would open up an important new field, and very probably facilitate the study of the human mind as a natural phenomenon.

Now from the point of view of modern physics, the world of classical physics, the world of particles in which Laplace believed, is an approach to the truth. I think we can now assert that if the real world consisted of such self-existent particles there would be no room in it for mind. Further if, as we saw reason to suspect, the world of the remote past was a world of particles, or very nearly so, life and mind were then impossible. It is very difficult to see how mind could either perceive a world like that, or act upon it. Descartes' theory is unsatisfactory. On the other hand, my own view is that it is just through the contradictions in the crude view of nature that we can see, if only in the dimmest way as yet, how it has been possible for life, and later for mind, to arise.

At the opposite pole to Newton's and Locke's conception of the world as consisting of moving particles, we have Plato's timeless world, consisting of unchanging ideas, such as those of squareness, or of good. Attempts are made from time to time by modern philosophers to revive a belief in Plato's ideal world. Whitehead, for example, has done so. These attempts lead to strange results. For clearly, if there is a timeless world, the events in the world of time cannot influence it. This would be all very well if the timeless world had no causal relation with nature. But this ceases to be true, in Whitehead's philosophy, when mind appears on the scene. A bit of water freezes. This is described as the "ingression" into the natural world of the eternal object "iciness." But iciness is not the cause of freezing. The cause is to be found in time. However, when a philosopher observes the pond he becomes aware of the existence of the eternal object which he calls iciness. He gives it a name which he writes down on paper, and hence, as the result of his apprehension of it, the visible world has been changed. Thus eternal objects cause happenings in the visible world, whilst the converse is not true. This seems to me an objectionable hypothesis. It is quite analogous to the theory that a mind can observe the world without acting upon it. If, however, the idea of iciness, so far from being eternal, only exists in the minds of philosophers, it is, for a thoroughgoing materialist, a part of nature, though with properties so unlike those of an ordinary material object that Plato's theory is quite a useful approximation to the truth. The less massive an object, the less definitely it is localized in time and space, and the more nearly it approximates to a Platonic idea. The more massive, the better is its approximation to a Newtonian particle, and it is in the sphere of astronomy that Newton's ideas have been most successful.

Whether or not such speculations as these prove to be anywhere near the truth, they show the immense gulf between dialectical materialism and the "vulgar" materialism of the nineteenth century. For T. H. Huxley, who did not even call himself a materialist, mind was an epiphenomenon of matter. Mental events were determined by physical events and did not determine them. The mind was a powerless spectator of the play of matter, as a nineteenth-century academic philosopher was a powerless spectator of the political and economic events of his time.

This mechanistic attitude is still alive today as behaviourism, the system of J. B. Watson, which ignores conscious processes completely. It appears to have some practical value for advertisers, and perhaps in other fields. This is not surprising, for any abstraction may help us to concentrate on a given isolated field of phenomena. But it is based on a faulty physiology, particularly of the ductless glands, and if only for this reason, is unlikely to be of a very lasting importance.

The distinction between Marxism and the more

idealistic types of academic philosophy, which generally have a more or less religious tinge, is of another kind. Such philosophies admit that mind can act upon the visible world, which indeed is often regarded as really of the nature of mind rather than matter. But because mind is logically and temporarily prior to matter, real progress becomes illusory. At best, the mind may get back to God, or may free itself from matter or from the illusion that matter is real. Good is taken as an absolute. The Marxist knows nothing of absolute good, but he can recognize the relatively better, and he can fight for it. "*Le mieux est l'ennemi du bien*" is a very profound saying.

The recent only academic philosopher in England whose system has any serious affinity with Marxism was Alexander. He tried to trace the evolution of being from space-time, through matter to life and mind, and beyond mind to a hitherto non-existent quality which he perhaps rather unfortunately called deity, and which will be related to mind as mind is to mere life. While I think that he largely ignored both historical facts and the dialectical character of progress, there are some aspects of his thought which a Marxist might well accept.

For Marxists the mind is not something whose natural powers are blunted by association with matter. It is a part of nature still evolving, and still very imperfect. But some at least of its imperfections are the symptoms of contradictions both within society and within itself which are the conditions for further progress. It has risen from the mud, not fallen from heaven, and it is destined to rise still farther. Such a philosophy is at least profoundly optimistic. It enables Marxists to carry on through defeat, terror, and persecution. Although it offers no future life for the individual, the belief in better future lives for the human race docs, as a matter of observation, give to many Marxists the same energy and confidence that the hope of personal immortality gave to the early Christians.

Marx and Engels were primarily occupied by considering the mind in society. They did not deny freedom of the will in the sense of suggesting either that your will is not your own, or that it is a mere feeling associated with such vital processes as muscular contraction. Our wills are very largely determined by society. Moreover, the resultant of many wills is different from any of them. This was pointed out by Rousseau in his discourse on the *volonté générale* and the *volonté de tous* in the *Contrat Social.*

Here is what Engels wrote:*

"In one point, however, the history of the development of society proves to be essentially different from that of nature. In nature—in so far as we ignore man's reactions upon nature—there are only blind unconscious agencies acting upon one another and out of whose interplay the general law comes into operation. Nothing of all that happens—whether in the innumerable apparent accidents observable upon the surface of things, or in the ultimate results which confirm the regularity underlying these accidents—is attained as a consciously desired aim. In the history of society, on the other hand, the actors are all endowed with consciousness, are men acting with deliberation or passion, working towards definite goals; nothing happens without a conscious purpose, without an intended aim. But this distinction, important as it is for historical investigation, particularly of single epochs and events, cannot alter the fact that the course of history is governed

* F., p. 58.

by inner general laws. For here, also, on the whole, in spite of the consciously desired aims of all individuals, accident apparently reigns on the surface. That which is willed happens but rarely; in the majority of instances the numerous desired ends cross and conflict with one another, or these ends themselves are from the outset incapable of realization or the means of attaining them are insufficient. Thus the conflict of innumerable individual wills and individual actions in the domain of history produces a state of affairs entirely analogous to that in the realm of unconscious nature. The ends of the actions are intended, but the results which actually follow from these actions are not intended; or when they do seem to correspond to the end intended, they ultimately have consequences quite other than those intended. Historical events thus appear on the whole to be likewise governed by chance. But where on the surface accident holds sway, there actually it is always governed by inner, hidden laws, and it is only a matter of discovering those laws."

From the standpoint of modern psychology this passage can be criticized as ignoring the importance of unconscious tendencies, such as repressed sadism and masochism which show themselves in war. It will be said that the "inner, hidden laws" are to be found in the individual unconscious rather than in society, and that conflict can only be prevented by a mass application of psycho-analytical principles.

But while admitting the importance of unconscious tendencies we may ask ourselves why they find different outlets in different societies. And as soon as we do so we are led back towards the Marxist position. Wars would not occur if no one got psychological satisfaction from them. But it is certain that war preparations would not be made on their present scale if they did not offer unique opportunities not merely to make profits, but, by disciplining the workers as soldiers under the orders of members of the ruling class, to perpetuate existing class distinctions.

The avoidance of war, like the avoidance of fires, is a practical problem. Houses would not burn if the air contained only 15 per cent of oxygen, if they were all built of ferroconcrete, if everyone was both good and careful, or if matches and fires were abolished. The problem is simply which of several possible (or impossible) preventive measures we should choose. And as the existing economic structure of society is in any case unstable, it seems reasonable to work for a stabler structure in which many of the causes of war will be abolished. This does not mean that the psychological approach is nonsense. It means that it is probably of secondary importance.

And meanwhile it is curious to note that psychologists who find the predominant cause of war in the internal stresses of individuals often appear to be blind to its causation by the internal stresses of society. Hitler may well be a neurotic. But it is only in particular circumstances that a great people chooses a neurotic of this kind as its leader. And it is a more practicable task to abolish these circumstances, particularly mass unemployment, than to see that no neurotics are available as leaders, even in neighbouring nations (for Hitler was of course born an Austrian).

Further there is evidence* that the abolition of unemployment and class antagonisms has a large effect in diminishing neurosis. As I have not had access to the

* *Soviet Russia Fights Neurosis,* by Frankwood Williams. (New York 1936.)

statistical evidence for this statement, I do not of course insist upon its truth. A rough calculation suggests that if we accepted the extreme claims of the psycho-analytical school, we should need a corps of some 50,000 psycho-analysts to keep England tolerably free from neurosis. The cost would be somewhat less than that of the navy, though of the same order, but even if enough men and women of the right type are available, a generation would be needed before they could be trained. And the menace of war is immediate.

So though I do not for a moment deny the claims of individual psychology, an investigation of the laws which govern the development of societies would seem to be of more immediate importance. In the last chapter I shall deal very briefly with some of the laws which, according to Marx, govern this development.

6. *Sociology*

I CONTINUE the quotation from Engels with whose consideration I closed the last chapter.

"But while in all earlier periods the investigation of these driving causes of history was almost impossible—on account of the complicated and concealed inter-connections between them and their effects—our present period has so far simplified these inter-connections that the riddle could be solved. Since the establishment of large-scale industry, i.e. at least since the peace of Europe in 1815, it has been no longer a secret to any man in England that the whole political struggle there has turned on the claims to supremacy of two classes: the landed aristocracy and the middle class."[1]

You will remember this was written just over fifty years ago, when the Conservative and Liberal Parties respectively represented these two classes, as the present National Government represents their combined interests.

"Since 1830, the working class, the proletariat, has been recognized in both countries as a third competitor for power. Conditions had become so simplified that one would have had to close one's eyes deliberately not to see in the fight of these three great classes and in the conflict of their interests the driving force of modern history—at least in the two most advanced countries.* But how did these classes come into existence? If it was possible at first glance still to ascribe the origin of the great, formerly feudal landed property—at least in the first instance—to political causes, to taking possession by force, this could no longer be done in regard to the bourgeoisie and the proletariat, it was a question in the first instance of economic interests, to the furtherance of which political power was intended to serve merely as a means."

You see that he denies the importance in the creation of a class, on the one hand of mere force, the mere seizure of power, in contradistinction with those people, including some anarchists, who say that property is theft; but, on the other hand, he stresses the importance of economic developments as compared with the political and legal developments, which he claims to be their mere image, as it were.

Now the most ambitious general account of history given by any Marxist so far has been Engels' book, *The Origin of the Family, Private Property, and the State,* which was published in 1884. Marx had intended to write it, but he died before this was possible. It is an account of the early transformations of human society; it is based to a very large extent on the work of three men. Bachofen was the first to establish the existence in early European society of traces of matriarchy, that is to say, to adduce evidence that in the past the mother,

* France and England.

rather than the father, was head of the family and determined the larger group of which the family was a part. McLennan investigated the society of the Australian natives, and in particular the division of their tribes into exogamous groups, within which marriage was forbidden. Above all Engels was influenced by the work of Morgan, who based a number of books, especially *Ancient Society*, on a study of the Iroquois Indians of New York State. He had been a member of one of their gentes, as he called them, and his work represents the beginning of the great modern American school of anthropology which has done so much to give us a detailed picture of how primitive societies actually work. On the basis of these data, and of his economic theories, which were to some extent shared by Morgan, Engels attempted a synthesis. The defects of that synthesis are now fairly obvious—I shall deal with some in a moment. We have on the other hand to consider its background. At the time when he wrote the official pre-history as taught in England was based partly on the book of Genesis and partly on Homer. We knew extremely little about the Stone Ages, and perhaps the most popular work on pre-history was Mr. Gladstone's book, *Juventus Mundi*, in which "he presents to his readers the Grecian chiefs of the heroic age as kings and princes, with the superadded qualities of gentlemen"! Engels was one of the numerous people who pointed out that the chiefs were very far from being gentlemen (as is fairly obvious to anyone who reads his Odyssey). But he also added that they were very far from corresponding to kings and princes in the states of the last two thousand years.

The alternatives to this official theory were almost

equally fantastic, for example, Rousseau's theory of the social contract, according to which at some time in the past, a number of people had got together and said, "Let's have a state"; and had arranged, like business men sitting round a table to amalgamate a number of firms, to give up certain rights in favour of others. The contemporary British writers were just making the discovery of the progress from status to contract as a fairly general feature of human society in the last few thousand years, and this was being put forward as a great historical principle. For Marx, it was undoubtedly one feature of social change, but it was a somewhat superficial feature, because it was legalistic. The question was not so much, "Why does a person now make a contract for such and such wages in return for his work when six hundred years ago he did not make such a contract?" but "Why has society so changed that this is necessary?" If Engels were writing today, he would have vastly more data, particularly on pre-historical technique, notably details of the paleolithic and neolithic stages of human development; and he would have also far more data on existing or recently extinct primitive societies. In consequence he would find it very much harder to draw the rather sharp lines which he did.

That shading out of sharply-drawn lines is an inevitable result of greater knowledge. If you compare the attempt to construct human pre-history of fifty years ago with the attempts to construct the pedigree of human evolution you find the same thing. A great many gaps have been filled in. Where Huxley and Haeckel could point to a relatively sharp gap between fish and amphibian, or between reptile and mammal, we now have a considerable number of fossil forms which cannot be assigned with great confidence to either, and constitute their link. It is harder than it was fifty years ago to give a concise account of the evolution of vertebrates. In spite of that, however, the principal conclusions at which Huxley and Haeckel arrived form a starting point for an account of evolution, and are still true. My own view is that the same holds good to a very large extent for the system laid down by Engels, though there are certain parts of his scheme which are obviously doubtful and a few which are probably false. He distinguishes several different stages of savagery, the highest being characterized by the use of the bow and arrow. These are followed by barbarism, with successive stages characterized by pottery, domestication, and metallurgy. After that comes civilization, with various stages, of which in Europe the most important are the antique or Graeco-Roman civilization based on slavery, feudalism, the capitalist system, and the socialist system, which was then in the future, but is now in one country in its very early stages. As Marx pointed out, these may shade off into one another to a considerable extent. There was generally no sharp break between feudalism and capitalism, though the change may, in certain cases, be abrupt. In all cases, social changes are thought to be brought about by changes in the means of production. Of course, under the word "production," Marx included transportation and other processes by which human labour increases the value of goods.

One of the interesting results of Marxism has been an attempt on a very large scale to investigate as far as possible the early history of the existing means of agricultural production. It has become clearer and clearer that the domestication of animals and plants was an enormously revolutionary step; and it was certain that we could get no very detailed picture of the progress of man from savagery to civilization unless we could get some idea of when and where those changes took place. That has been accomplished to a considerable extent in the case of the more important agricultural plants by the work of Vavilov and his colleagues in the U.S.S.R. Vavilov, for example, investigated the question of where wheat had originated. It was no good going to tradition. There you found legends of mythical inventors of agriculture like Triptolemus in a number of different countries. It was little good even examining the wild plants which are closely related to wheat, as de Candolle had done, because some may have been spread as weeds in the wheat crop, others exterminated by agriculture, and so on. Vavilov's main method was of a different character, which I can illustrate by the case of maize. Maize is well known to have been brought to the Old World from Central and North America. If we examine the number of types of maize in any given country, we shall find that almost everywhere in the New World there are a great many more different types of maize than in the Old World, particularly so among the native populations which cultivate them. We find plants with purple seeds, and a great many morphological types which are abnormalities from the point of view of Europeans who have only used a limited number of the varieties available. We find a centre of diversity for maize in and about Mexico; and although there is nowhere wild maize, we find there a grass called teosinte which will cross with maize, and very probably represents its ancestral wild form.

Applying the same principles to wheat, Vavilov looked for the centres of diversity of different types of wheat. For there are a number of different types which will not interbreed with one another except with great difficulty, and which have originated in different places. He found, for example, the centre of diversity of the group of wheats with twenty-eight chromosomes, of which the most important is probably macaroni wheat, round the Eastern Mediterranean; but, on the other hand, the centre of diversity for bread wheats with forty-two chromosomes is somewhere in the neighbourhood of Afghanistan. To give an idea of the data on which his conclusions are based, he found only twenty sub-species of bread wheats in the whole of Europe, forty-five to fifty in Persia, and something like seventy in Afghanistan. If you go in the other direction from Afghanistan into India, you find a type of bread wheat there, *Triticum sphaerococcum*, which is quite different from anything found in Europe, and which, like the European wheats, may well be derived from some of those in Afghanistan.

On the basis of analyses of this type, he came to the conclusion that plants were domesticated, not by any means all over the world, and not, as has been commonly supposed, in the great river valleys like those of the Nile, the Euphrates, and the Indus, but in certain mountainous regions, from which they have been later brought down into the plains.

Some of his work may turn out to be fallacious; it will certainly have to be revised in detail. Nevertheless it is extremely interesting as an example of the kind of work on which a biologist embarks under the influence of Marxist ideas, and in a society where those ideas are dominant.

There are several other points to be considered with regard to the origins of agriculture. One point, which is so simple that I think I can make it in the space at my disposal, is this. We find fairly primitive peoples cultivating wheat, potatoes, or other food plants which are at least a very great advance on anything which is found in the wild state; and it is commonly thought that those plants can only be the result of a prolonged and conscious effort in the way of selective breeding, such as has considerably improved the cereals and potatoes of Europe in the last two centuries. I believe that this is quite incorrect. It is a typical attempt to ascribe a process which goes on unconsciously to the deliberate will of the people concerned. What has actually happened, I think, is this. As you domesticate a plant such as wheat, as soon as you start gathering the seed and sowing it you will automatically select the most fertile plants. The process is more rapid in the case of self-fertilizing plants like wheat than of cross-fertilized plants such as maize, because a maize plant may be very successful as a father, on account of producing large amounts of pollen, and yet produce relatively few or small seeds, and thus its genes may survive under domestication. However, even with maize the process is fairly quick. This process will give a cereal of fairly high quality within a very few centuries, or even a few human generations.

This principle only applies to plants in which we actually use for our own purposes those parts with which it reproduces. For example, if you go to the ordinary seedsman and buy what purports to be meadow grass, you will probably find that when you sow your seed, you get something a good deal less satisfactory than the grass which you will find in a good meadow. The reason is quite simple. The seedsman takes plants from a meadow which he thinks likely to be good doers. He gathers the seeds for a number of generations, and finally by this automatic process obtains meadow grass plants producing a fine crop of grass seed, but which produce relatively few and small leaves, and will not stand up to being grazed by cows, horses, and sheep. Professor Stapledon, of Aberystwyth, by carefully selecting individual plants, has been able to get over the unconscious effect of selection, which in the case of meadow grass is as harmful as it is beneficial in the case of wheat. These are examples of the way in which biology is being used to illuminate the early stages in human history.

The Russian writers now distinguish several different stages in the exploitation of a domestic animal. The earliest stage is the one in which it is exploited primarily for its meat and skin. There is a later stage where it is exploited for produce which does not involve killing it—its milk, its wool, or its power. In England the rabbit is just passing into the second stage with the breeding of Angora rabbits which are periodically shorn for their wool, and are therefore of value during their lifetime, instead of being of no value until they are killed, like ordinary rabbits.

In each stage of social development based on a particular productive technique, there is a characteristic organization of the family, of property, and of political aggregates. On the basis of the classificatory system of kinship, Engels believed, with many contemporaries, that there had been a stage of group marriage. If in a given society a man uses the same term for his own wife and for a group of other women, this was thought to point back to an earlier stage when he would have been one of the husbands of all these women. The existence of this stage is denied by many recent anthropologists. It is possible that Westermarck has gone a little too far in his criticism of the group-marriage theory. One cannot too lightly dismiss the evidence from such societies as that of the Lepchas in Sikkim, where a man has the right of access to the wives of his elder brothers, but a great many unions which would be perfectly legitimate in Europe are barred as incestuous.

The whole discussion of this question has been revolutionized by the alleged discovery that some primitive men do not understand the physiology of paternity; and it is therefore to some extent true that the whole question is in the melting-pot. Nevertheless, in the next stage, which undoubtedly exists in many lands, we find an organization of society which is very strange to us. Society is largely based on the gens, as Engels calls it, which is exogamous in the sense that a woman must marry a man belonging to a different gens. There are from two to eight of these gentes in a tribe, the membership of a gens is usually based on maternal inheritance; and such little property as there is, is handed down in the maternal line.

The succeeding stage undoubtedly exists in what Engels called the pairing family, a rather loose monogamy in which the possibilities of marriage are governed by exogamous rules. Property is more or less communal, even where there are families of one man and one woman. Many such families frequently live in a communal household. Then follows the fourth stage, characterized by monogamy, at any rate for women, with a rigid marriage system, male supremacy, and inheritance in the male line. With that particular change, which Engels is inclined to attribute to the domestication of large numbers of animals, you pass over to the patriarchal family so well described in the book of Genesis, and we have left the stage of primitive communism. This change, though probably an inevitable step on the road to civilization, had its disadvantages. Here is what Engels writes on the subject:*

"How wonderful this gentile constitution is in all its natural simplicity! No soldiers, gendarmes and policemen, no nobility, kings, regents, prefects, or judges, no prisons, no lawsuits, and still affairs run smoothly. All quarrels and disputes are settled by the entire community involved in them, either the gens or the tribe or the various gentes among themselves. Only in very rare cases the blood revenge is threatened as an extreme measure. Our capital punishment is simply a civilized form of it, afflicted with all the advantages and drawbacks of civilization. Not a vestige of our cumbersome and intricate system of administration is needed, although there are more public affairs to be settled than nowadays; the communistic household is shared by a number of families, the land belongs to the tribe, only the gardens are temporarily assigned to the households."

He goes on to analyse in detail evidence that a change of this kind happened in late European pre-history, particularly in the early history of Attica, of which

* O.F., p. 117.

Athens was the principal town.* Having described the gradual change-over in property relations, particularly due to the accumulation of cattle in the hands of individual families where they were handed down in the paternal and not the maternal line, he goes on to say:†

"That is one side of the question. We must not overlook, however, that this organization was doomed. It did not pass beyond the tribe. The league of tribes marked the beginning of its downfall, as we shall see, and as the attempts of the Iroquois at subjugating others showed. Whatever went beyond the tribe, went outside of gentilism. Where no direct peace treaty existed, there war reigned from tribe to tribe. And this was carried on with the particular cruelty that distinguishes man from other animals, and that was modified later on simply by self-interest. The gentile constitution in its most flourishing time, such as we saw it in America, presupposed a very undeveloped state of production, hence a population thinly scattered over a wide area. Man was almost completely dominated by nature, a strange and incomprehensible riddle to him. His simple religious convictions clearly reflect this. The tribe remained the boundary line for man, as well in regard to himself as to strangers outside. The gens, the tribe, and their institutions were holy and inviolate. They were a superior power instituted by nature, and the feelings, thoughts, and actions of the individual remained unconditionally subject to them. Commanding as the people of this epoch appear to us, nothing distinguishes one from another. They are still attached, as Marx has it, to the navel string of the primordial community. The power of these natural and spontaneous communities had to be broken and it was. But it was done by influences that from the very beginning bear the mark of degradation, of a downfall from the simple moral grandeur of the old gentile society. The new system

* See G. Thompson on "The Social Origins of Greek Tragedy," *Modern Quarterly*, Vol. 1.

† O.F., p. 118.

of classes is inaugurated by the meanest impulses: vulgar covetousness, brutal lust, sordid avarice, selfish robbery of common wealth. The old gentile society without classes is undermined and brought to fall by the most contemptible means: theft, violence, cunning, treason. . . .

"Only one thing was missing: an institution that not only secured the newly acquired property of private individuals against the communistic traditions of the gens, that not only declared as sacred the formerly so despised private property and represented the protection of this sacred property as the highest purpose of human society, but that also stamped the gradually developing new forms of acquiring property, of constantly increasing wealth, with the universal sanction of society. An institution that lent the character of perpetuity not only to the newly rising division into classes, but also to the right of the possessing classes to exploit and rule the non-possessing classes. And this institution was found. The State arose."

Now this is a rather surprising conclusion from the point of view of people who have no great acquaintance with Marxism. They expect to find the State appearing in socialist writings as something grand and noble, destined to expand, and ultimately to take over all private property. They are surprised to find the State regarded as something whose origin was an evil, if a necessary evil. Engels goes on to describe the development of Graeco-Roman society, and here again one might expect him to fulminate against slavery, which is always easy to do if you belong to a society in which it has been abolished. You will find* that he regarded it as an inevitable development:

"It is clear that so long as human labour was still so little productive that it provided but a small surplus over and

* A.D., p. 206.

above the necessary means of subsistence, any increase of the productive forces, extension of trade, development of the State and of law, or beginning of art and science, was only possible by means of a greater division of labour. And the necessary basis for this was the great division of labour between the masses discharging simple manual labour and the few privileged persons directing labour, conducting trade and public affairs, and, at a later stage, occupying themselves with art and science. The simplest and most natural form of this division of labour was in fact slavery. In the historical conditions of the ancient world, and particularly of Greece, the advance to a society based on class antagonisms could only be accomplished in the form of slavery. This was an advance even for the slaves: the prisoners of war, from whom the mass of the slaves was recruited, now at least kept their lives, instead of being killed as they had been before, or even roasted, as at a still earlier period."[1]

This passage is, of course, somewhat out of date. Modern imperialistic wars are waged for the possession of mineral resources or strategic advantages, rather than to exploit the labour power of the conquered. In consequence the technique of war has now somewhat altered, and the vanquished civilian population is once more massacred, as in Spain and China, or even roasted, as in Abyssinia. Engels may perhaps be forgiven if he did not foresee this highly characteristic development of imperialism.

He goes on to point out that subordination of the workers, although it was necessary at the early stage of civilization, is now no longer necessary:*

"Only the immense increase of the productive forces attained through large-scale industry made it possible to

* A.D., p. 207.

distribute labour over all members of society without exception, and thereby to limit the labour time of each individual member to such an extent that all have enough free time left to take part in the general—both theoretical and practical—affairs of society. It is only now, therefore, that any ruling and exploiting class has become superfluous and even a hindrance to social development, and it is only now, too, that it will be inexorably abolished—however much it may be in possession of the 'direct force.' "

Throughout, Engels strongly combatted the view that property is theft, and Marx traced the development of our existing property system, mainly, though not wholly, by means which were legal and even admirable in their time. In particular, at a time when under the influence of the rather extreme Protestantism expressed by such writers as Macaulay the Middle Ages were despised by English writers, Marx and Engels took the view that they were a period of fairly steady growth of industry and particularly of industrial skill among the free workers; and that it was on account of this development of skilled labour, a development which had been impossible in the slave-run society of Greece and Rome, that modern industrialism became possible.

Marx particularly described the rise of capitalism and its internal contradictions in *Capital.* I am not qualified to do so. It is far too vast a subject for me to attempt. I would merely point out that his most important original contribution to economics is undoubtedly the theory of surplus value, the theory that the worker sells his labour power, for which he receives a wage; and that the surplus value which he does not receive in his wage is the difference between that wage and the value which he has added to the commodities by his work. It is, however, of the utmost importance to realize that though Marx and Engels thought capitalism was unjust, their reason for believing that it would come to an end within a relatively short time was not because it was unjust, but because it was not working.

> "But just as, at a definite stage of its development, manufacture came into conflict with the feudal order of production, so now big industry has already come into conflict with the bourgeois order of production established in its place. Tied down by this order, by the narrow limits of the capitalist mode of production, big industry produces on the one hand an ever increasing proletarianization of the great mass of the people, and on the other hand an ever greater mass of unsaleable products. Over-production and mass misery, each the cause of the other—that is the absurd contradiction which is its outcome, and which of necessity calls for the liberation of the productive forces by means of a change in the mode of production."*

I shall not elaborate this point in any detail. There is much in the above paragraph which is generally admitted now, though it was not admitted fifty years ago, except by Marxists. What I wish to emphasize is the relationship of Marxism to the State. The Utopian socialists of the early nineteenth century claimed to demonstrate that socialism would work better than capitalism, and they hoped to persuade the ruling class to make the necessary change. Owen, in particular, who had been an extremely successful factory manager, was actually surprised when the best people dropped him like a hot brick as soon as he became a convert to socialism. The Marxist criticism of such writers is simply that they do not sufficiently realize how far the prevailing ideas are determined, not

* F., p. 62.

on a basis of pure reason, but on an economic basis. For example, today Mr. Keynes would like to persuade the capitalists to be contented with a greatly lowered rate of interest; and I have very little doubt that if Mr. Keynes were able to carry out this and other reforms, capitalism could be made to last for a considerable time. From the Marxist point of view one of the main defects in Mr. Keynes' writings is that he never explains why the capitalists do not follow his advice and are not likely to do so; why, for example, the British Government at the time that the Versailles Treaty was drafted did not follow his advice. A Marxist would give good reasons why his advice could not then, and will not now, be followed. It is just for that reason that Marxian socialists claim to be scientific and not Utopian. They recognize the type of historical process by which changes are brought about. They are not brought about by a number of people sitting down together and saying, "Let's have a nice plan which will make us all happy!" That idea corresponds to the theory of Rousseau, which Marxism claims is an unhistoric theory. Actually, changes are produced by struggle as much as by agreement, and it is necessary to recognize this fact, even when you are doing your best to introduce some reason into the process.

Marx and Engels recognized that the transition to Socialism could only be brought about by the efforts of the workers. Engels certainly reached this view independently of Marx, and possibly before Marx. It is very instructive to compare his book, *The Condition of the Working Class in England*, written in 1845, with the writings of his contemporary, Charles Dickens. Dickens had a first-hand knowledge of these conditions. He described them with burning indignation and in great detail. But his attitude to them was one of pity rather than hope. Engels saw the misery and degradation of the workers, but he saw through it. Dickens never suggested that if they were to be saved they must save themselves. Engels saw that this was not only desirable but inevitable.

Today, on the right, so to speak, of the Marxists, are the non-Marxist socialists, including most of the Labour Party in this country, who hope to take over the existing State and make it the owner of the most important means of production, for example, of the railways, the land, and the coalmines, at an early date. On the left of Marxism you have in Southern Europe, though not in large numbers in England, the anarchists, who wish to smash the existing State because it is the bulwark of capitalism. They think that if this is done, capitalism, being based on force, will die without a struggle.

The Marxists differ fundamentally from both. The Marxist point of view is based on a particular theory of the origin of the State, the theory that it originated with the division of mankind into economic classes and that it is determined by the class struggle, and not the other way round. From the conservative point of view, the State exists to keep order. From the radical point of view, the State exists to perpetuate existing distinctions. Whichever way you look at it, Marxists claim that it is an expression of the class struggle.

Now the Marxists, like the anarchists, are out to abolish the State. Marx agreed to a surprising extent with the extreme nineteenth-century liberals, and with philosophers like Herbert Spencer, when he characterized the State as a "parasitic excrescence" to be "amputated." For example, he wrote: "The [Paris] Commune realized the slogan of all bourgeois revolutions, cheap government, by eliminating the two largest items of expenditure, the army and the bureaucracy." He wanted to abolish what he regarded as the overgrown State of his own time. On this point, if on few others, he was, as I say, in agreement with the rather old-fashioned Manchester School of liberals. Further, he wrote: "The working class cannot simply lay hold of the ready-made State machinery and set it going for its own ends." Here it is important to realize that the theory of the Communist Manifesto, published in 1847, was modified in the light of the experience of the Paris Commune of 1871. It was Blanqui who coined the rather cumbrous phrase "the dictatorship of the proletariat" to express the type of organization which, according to Marxist theory, must take over for a time the functions, or some of the functions, of the existing State in the early stages of socialism.

I may add that the phrase "dictatorship of the proletariat" may be adequately translated into ancient Greek as *δημοκρατία* (democratia). This word is derived from *δῆμος* (demos) which, according to Liddell and Scott's Lexicon, means the people, or commons, as opposed to the wealthy, the fat, the powerful, the few, etc., and *κράτος* (kratos) meaning strength, power, rule, or victory. A Greek democracy was governed by an *ἐκκλησία* (ekklesia), which was a moot or soviet of all the citizens in a town (though not, of course, of the slaves). This body used its power with the utmost vigour, and often with little regard for law.

I do not know whether Plato and Aristotle would have

been more surprised to find the word democracy applied to a State like Great Britain with a hereditary monarch and peerage, and no direct rule by popular assemblies, or to find such undemocratic bodies as the Roman and Anglican Churches denoted by the word ecclesia (*église*). They might, however, have described the Soviet Union as an ecclesiastical organization!

At any rate, the Greeks meant by a democracy something not unlike the dictatorship of the proletariat as it appears in Marxist writings round about the year 1871. It is important to note that Marx regarded it as a temporary measure. As the State was only the expression of class struggle it would be bound to wither away when the class struggle was over, even if in the concluding stages of the struggle a very powerful State organization was necessary.

One of the Marxist criticisms of the anarchist point of view is that the anarchists put the cart before the horse. Because the State is the expression of class struggle, you are bound to have a State going on so long as the class struggle continues. Lenin, in his remarkable book, *State and Revolution*, developed this doctrine, which is also expressed in Marx's *Civil War in France*, published in 1872, and in the revised form of the Communist Manifesto. Lenin's other principal development of Marxism was, I think, the attempt to bring the account of capitalism given by Marx up to date in view of the more recent developments of monopoly and finance capital throughout the world. On the whole, as might be expected, he stresses the later stage of Marx's thought, after 1871, at the expense of the earlier stage.

There are those who claim that Lenin has hopelessly

distorted Marxism; and no doubt he left it somewhat different from what he found it. But it is difficult and in my opinion impossible to point to any particular respect in which he actually went against Marx; and many people regard his development of Marxism as entirely legitimate, and in the spirit of a philosophy which is by its very nature a method rather than a doctrine.

I know that the Marxist interpretation of our present troubles, economic, political, and ideological, is at any rate something which makes for hope and faith in the future. I also regard it as substantially true. Whether it is true or not, it certainly produces the grim optimism which we find in St. John's Revelation, where he said: "The Devil is come down amongst you having great wrath, because he knoweth that he hath but a short time." That is, roughly speaking, the Marxist view of Fascism. It is important when people criticize Marxism as apocalyptical to remember that from the point of view of the Christian apostles the world as they knew it, the world based on slavery, was doomed to come to an end within a lifetime, or at most a few lifetimes, of their writing it down. It did come to an end—though not in the way in which they believed that it would. There was an immense social transformation, and an immense social transformation on the whole for the better.

What is more, St. John, though he wrote in mythological language, and was ignorant of a great many facts, had a better grip of the world situation than his contemporary historian Tacitus. Tacitus wrote from the standpoint of the ruling class. He knew which senators had been executed by a given emperor, and which general had commanded a given army. But St. John wrote from the standpoint of the oppressed masses, and he knew that the world in which he wrote was going to end in blood and fire. For this reason I can quote St. John in support of Marxism without being afraid of the criticism that his prophecies have not been fulfilled in detail.

We Marxists believe that a very big social change is inevitable in this country, but we believe that it is still possible, if that change is intelligently guided, that this country may avoid not only foreign war, but civil war. We believe, however, that the intelligent method is only going to be available if people will see the existing crisis against its historical background, if they will see that the immense growth of industry has produced a strain on human organization as great as that produced by the invention of domestication some ten thousand years ago. We believe that in so far as we can rationalize these changes, in so far as we can bring them about not by crude violence but by the application of a definite theory, we are likely to avoid some of the evils which were brought about by the earlier changes in the methods of production. We believe, then, that change is inevitable. It is for us to make it as rational, as human, and as quick as possible.

Index

Index

Index

Index

For Product Safety Concerns and Information please contact our EU representative GPSR@taylorandfrancis.com Taylor & Francis Verlag GmbH, Kaufingerstraße 24, 80331 München, Germany